Coordinated Energy Management for Low-Carbon Power Systems

Wei Lin · Juan Yu · Suhan Zhang

Coordinated Energy Management for Low-Carbon Power Systems

From a Resource Aggregation Perspective

Science Press
Beijing

Springer

Wei Lin
Chongqing University
Chongqing, China

Juan Yu
Chongqing University
Chongqing, China

Suhan Zhang
University of Sheffield
Sheffield, UK

ISBN 978-981-96-7532-6 ISBN 978-981-96-7533-3 (eBook)
https://doi.org/10.1007/978-981-96-7533-3

Jointly published with Science Press.
The print edition is not for sale in China mainland. Customers from China mainland please order the print book from: Science Press.

This Springer imprint is published by the registered company Springer Nature Singapore Pte Ltd.
The registered company address is: 152 Beach Road, #21-01/04 Gateway East, Singapore 189721, Singapore

Preface

Numerous countries have adopted national strategies for low-carbon power systems to resolve the environmental problems stemming from fossil fuel combustion. This drives the fast development of global renewable installations while the resultant volatility and uncertainties of renewable integration introduce challenges in maintaining the energy balance in the operations and planning of power systems. These challenges can be resolved by sufficient response and regulation abilities in the low-carbon power system, and thus necessitate coordinated energy management to maintain the energy balance in the operations and planning of the low-carbon power system. Typically, such coordinated energy management can occur in terms of the transmission level, distribution level, and interactions with heterogeneous energy flows.

This book provides systematical exhibitions of grabbing coordination opportunities for energy management in low-carbon power systems in terms of the transmission level, distribution level, and interactions with heterogeneous energy flows. Particularly, this book focuses on coordinated energy management for low-carbon power systems from a resource aggregation perspective, since the resource aggregation manner stands out for its privacy protection, and play-and-plug applicability in current power industries with superior coordination effectiveness.

This book consists of the following six chapters:

Chapter 1 introduces the necessity and opportunities of coordinated energy management with specific attention to the transmission level, distribution level, and interactions with heterogeneous energy flows. Additionally, the basic framework of resource aggregation-based coordinated energy management is outlined.

Chapter 2 introduces the resource aggregation methods for low-carbon power systems at the transmission level, together with their improvements in computational efficiency when facing various challenging issues (including the large scale of one transmission sub-systems, multiple border buses, and multiple time periods).

Chapter 3 introduces the resource aggregation methods for low-carbon power systems at the distribution level, which can be further categorized into node-level resource aggregation and system-level resource aggregation based on the implementation procedure.

Chapter 4 introduces the resource aggregation methods around interactions with heterogeneous energy flows, including gas systems and heating systems.

Chapter 5 introduces resource aggregation under renewable uncertainties which can be delineated by either probabilistic distribution functions or intervals. Additionally, the uncertainty impacts on the resource aggregation results are discussed.

Chapter 6 exhibits four application examples of the previous resource aggregation methods in the operations and planning of low-carbon power systems.

Finally, we would like to express our sincere appreciation to everyone who helped in making this book a reality. Our special thanks go to Prof. Wenyuan Li (Chongqing University, Chongqing, China), Prof. C. Y. Chung (The Hong Kong Polytechnic University of Hong Kong, Hong Kong SAR, China), Prof. Zhifang Yang (Chongqing University, Chongqing, China), Prof. Yi Wang (The University of Hong Kong, Hong Kong SAR, China), Prof. Changhong Zhao (The Chinese University of Hong Kong, Hong Kong SAR, China), Dr. Fei Feng (State University of New York Maritime College, New York, USA), Dr. Jianghua Wu (Shenzhen Research Institute of Big Data, Shenzhen China), Mr. Chenxi Wang (The University of Hong Kong, Hong Kong SAR, China), Mr. Linze Yang (Chongqing University, Chongqing, China), Mr. Qiyuan Li (Chongqing University, Chongqing, China), Mr. Fusu Li (Shanghai Jiao Tong University, Shanghai, China).

Hong Kong SAR, China
November 2024

Wei Lin

Contents

Abbreviations

AE	Algebraic equation
ATC	Available transfer capability
CB	Coal-fired boiler
CC	Combustion chamber
CDF	Cumulative distribution function
CHP	Combined heat and power
DER	Distributed energy resource
DG	Distributed generator
ED	Economic dispatch
EENS	Expected energy not supplied
ES	Energy storage
EV	Electric vehicle
GT	Gas-fired generator
IES	Integrated energy system
KKT	Karush-Kuhn-Tucker
LP	Linear programming
MFCR	Maximum flexibility capacity ratio
MILP	Mixed-integer linear programming
MINLP	Mixed-integer nonlinear programming
MLE	Maximum likelihood estimation
M-Sum	Minkowski Sum
NLP	Nonliear programming
ODAE	Ordinary differential-algebraic equation
PCC	Point of common coupling
PDAE	Partial differential-algebraic equation
PDF	Probabilistic distribution function
PTDF	Power transfer distribution factor
PV	Photovoltaic
QP	Quadratic programming
RODP	Reliability-oriented distribution planning
SoC	State of charge

ST	Steam turbine
STE	Steam-water heat exchanger
TSCH	Traverse-simulation-based concave hull

Chapter 1
Introduction

This chapter provides a comprehensive overview of coordinated energy management for low-carbon power systems from a resource aggregation perspective. First, this chapter outlines the background of coordinated energy management, through emphasizing its role in maintaining energy balance in low-carbon power systems dominated by renewable energy sources. Next, this chapter summarizes the opportunities for coordinated energy management for low-carbon power systems, categorized as transmission and distribution levels and interactions with heterogeneous energy flows. Finally, this chapter briefly introduces how to grab the opportunities from a resource aggregation perspective, including the definition, necessity, and key technological aspects of resource aggregation.

1.1 Energy Balance Requirements in the Low-Carbon Power System

In recent decades, the burning of fossil fuels has contributed to significant carbon emissions and thus resulted in a series of environmental problems (e.g., air pollution and the greenhouse effect) [1]. Since carbon emissions from power sectors share a considerable proportion of total emissions,[1] numerous countries have adopted national strategies for low-carbon power systems by promoting renewable integration, such as *Carbon peak and Carbon neutrality* in China [2], *Clean Energy for America Act* in the US [3], and *Germany's Renewable Energy Sources Act* in Europe [4]. Consequently, global renewable installations have been developing at an unprecedented speed, particularly driven by the falling installation costs and advancements in integration technologies; see Fig. 1.1 which reveals the fast development of global renewable installations from 2000 to 2022.

---

[1] International Energy Agency reported carbon emissions from power sectors neared 14.65 Gt in 2022, approximately occupying 38% of total emissions [1].

W. Lin et al., *Coordinated Energy Management for Low-Carbon Power Systems*,
https://doi.org/10.1007/978-981-96-7533-3_1

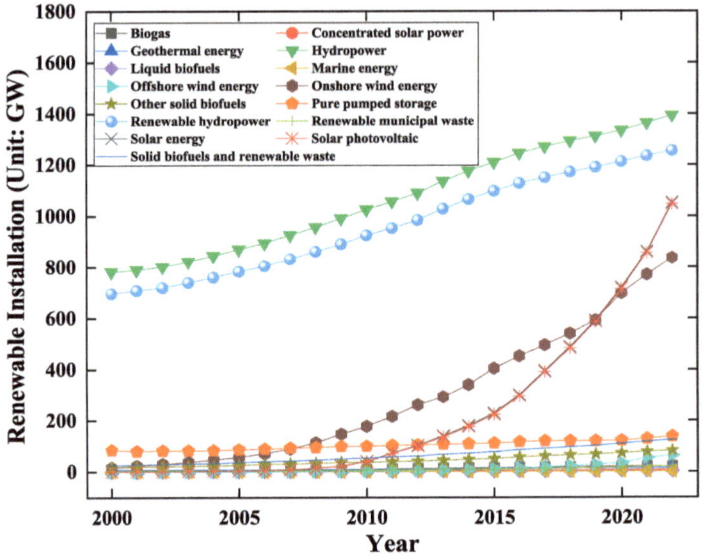

Fig. 1.1 Global renewable installations from 2000 to 2022 (data from [5])

The fast development of global renewable installations has effectively alleviated the environmental problems stemming from fossil fuel combustion, but has introduced challenges in maintaining the energy balance in the operations and planning of power systems [6]. For example, China's power system has been required to fully accommodate renewable integration for a long time. With the inherent volatility of renewable integration, more response and regulation abilities are required to ensure the energy balance in the operations and planning of China's power system [7]. Moreover, despite various prediction methods proposed for forecasting renewable integration, achieving accurate prediction remains challenging at this stage [8]; see Fig. 1.2 which reveals prediction errors among five popular prediction methods. Such prediction errors imply uncertainties of renewable integration; thus, more response and regulation abilities are additionally required to ensure the energy balance in the operations and planning of China's power system.

From the above, it can be found that the volatility and uncertainties of renewable integration require sufficient response and regulation abilities in the low-carbon power system. Such response and regulation abilities can be provided by investing new dispatchable resources (e.g., new thermal plants for regulation services) at the expense of the resultant increasing investment costs of energy supply [9]. Alternatively, coordinated energy management naturally offers a cost-effective solution by optimizing existing dispatchable resources during electricity generation, transfers, and load consumption. This helps to maximize the utilization of current dispatchable resources without additional investment costs.

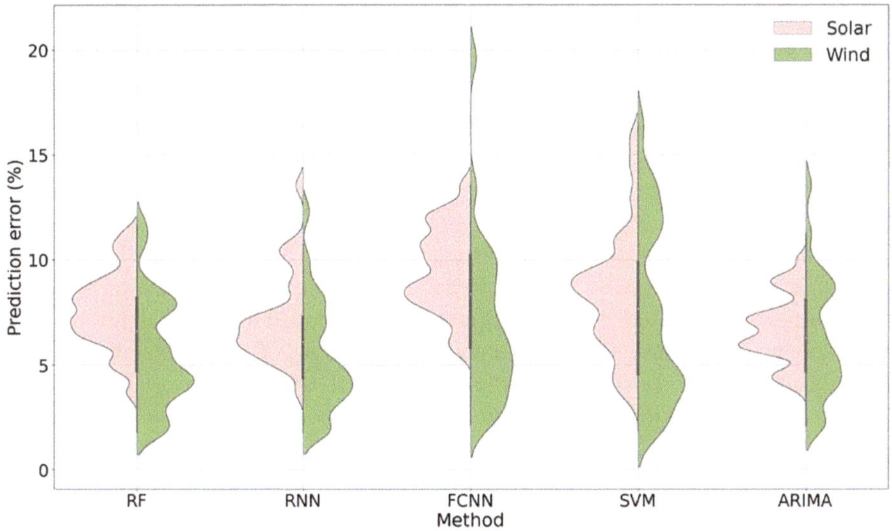

Fig. 1.2 Prediction error distribution across 30 provinces in China obtained by different methods, reprinted from [8], open access. The method to fit the density function as a ground truth is Kernel Density Estimation. RF, RNN, FCNN, SVM, and ARIMA indicate random forest, recurrent neural network, fully-connected neural network, support vector machine, and autoregressive integrated moving average, respectively. The smoothed curve in the left and right parts represents the prediction error density function across 30 provinces of solar and wind energy, respectively. The short black line in the middle of each shape is the median value of the data distribution, which visualizes the central tendency of the data distribution of each method

1.2 Opportunities of Coordinated Energy Management for the Low-Carbon Power System

The low-carbon power system encompasses transmission and distribution sub-systems within the electricity side and externally interacts with heterogeneous energy flows (e.g., heat flows and gas flows) [10]. Consequently, the dispatchable resources for the low-carbon power system are sourced from transmission and distribution sub-systems and their interactions with heterogeneous energy systems, necessitating coordination. In this context, this subsection outlines the opportunities of coordinated energy management for the low-carbon power system, through a taxonomy in terms of the transmission level, the distribution level, and interactions with heterogeneous energy flows.

1.2.1 Coordination Opportunities at the Transmission Level

Transmission sub-systems, which are crucial for transferring power from generation to load sides, are interconnected via tie-lines to enable power exchanges across different spatial regions. In the low-carbon power system, transmission sub-systems serve as transfer stations for distributing renewable integration across broader spatial regions, facilitating electricity generation resources to be optimized at the transmission level using tie-line power flow exchanges. For example, in mainland China, renewable energy sources such as wind and solar are abundant in western regions but loads are concentrated in eastern regions. Tie-line power flow exchanges facilitate renewable accommodation for such uneven distributions of renewable energies and loads through initiatives such as *West-to-East Electricity Transmission* project [11]. Another example is Germany's power system that is interconnected with another nine countries through 59 cross-border tie-lines [12]. Through tie-line power flow exchanges among different countries to manage the volatility and uncertainties of renewable integration, Germany's power system ensures its safe operations with high renewable penetration.

In summary, coordinated energy management at the transmission level can be achieved by tie-line power flow exchanges to coordinate electricity generation resources among different transmission sub-systems internally within the electricity side.

1.2.2 Coordination Opportunities at the Distribution Level

The traditional role of one distribution sub-system involves receiving electricity from its interconnected transmission counterpart at the *point of common coupling* (PCC) and subsequently distributing electricity to load customers. In other words, traditional distribution sub-systems are end-users observed from transmission sub-systems. However, this role is currently evolving with the integration of *distributed energy resources* (DERs) into distribution sub-systems [13]. DERs are small-scale generation/load resources in distribution sub-systems, such as rooftop *photovoltaic* (PV), *distributed generators* (DGs), and flexible loads (e.g., *electric vehicles* (EVs)). Consequently, the integration of DERs into distribution sub-systems brings the following two changes for the evolution from end-users to prosumers.

First, load customers in distribution sub-systems can be satisfied by either DERs (e.g., rooftop PV and DGs) or electricity transferred from one interconnected transmission sub-system. Moreover, electricity generated by DERs can be inversely transferred through the PCC to one interconnected transmission sub-system if necessary to address energy imbalance at the transmission level.

Second, flexible loads in distribution sub-systems can adjust their energy demand to support the energy balance in the low-carbon power system. For example, EVs can

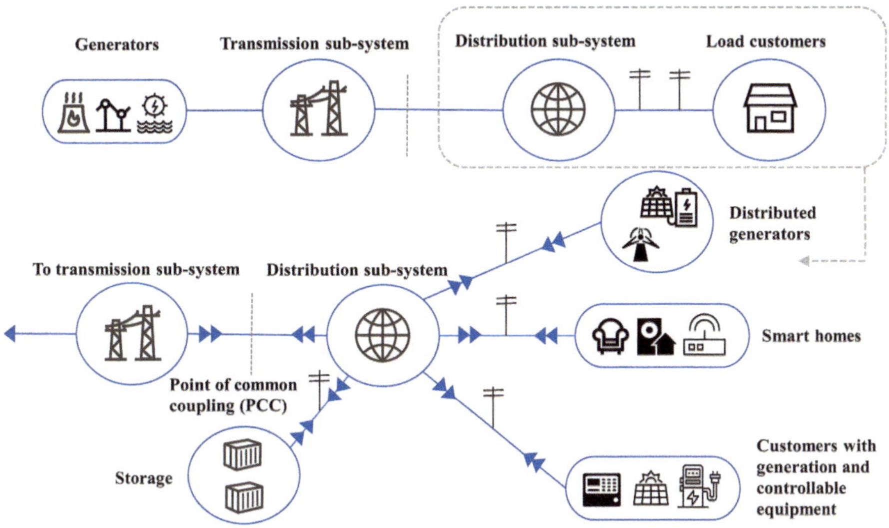

Fig. 1.3 Coordination between one transmission sub-system and its interconnected distribution sub-system, reprinted from [13], open access

mitigate their charging power when renewable integration in the low-carbon power system decreases, and vice versa.

Consequently, the PCC power flow exchanges between one transmission sub-system and its interconnected distribution sub-systems are no longer a single direction, but they can be bi-directional by optimizing DERs in distribution sub-systems, as shown in Fig. 1.3. In other words, coordinated energy management at the distribution level can be achieved by the PCC power flow exchanges between one transmission sub-system and its interconnected distribution sub-systems within the electricity side.

1.2.3 Coordination Opportunities by Interactions with Heterogeneous Energy Flows

In addition to the interconnections among different sub-systems within the electricity side (i.e., transmission and distribution sub-systems), the interactions of the electricity side with heterogeneous energy flows present another promising avenue for facilitating renewable accommodations in the low-carbon power system. As illustrated in Fig. 1.4, the interactions with heterogeneous energy flows can promote renewable accommodations from the electricity supply and consumption, as demonstrated below.

Regarding the electricity supply, surplus renewable energy can be converted to alternative forms of energy, stored, and reconverted into electricity when needed to

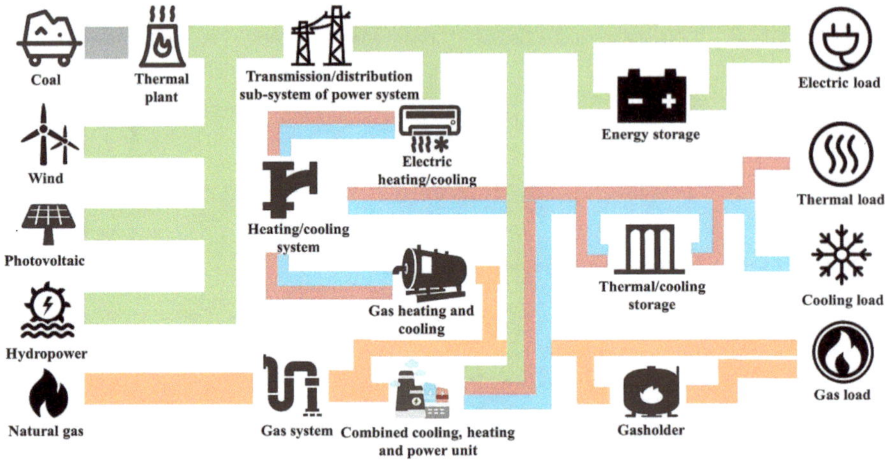

Fig. 1.4 Interactions of the electricity side with heterogeneous energy flows

alleviate energy imbalance on the electricity side. This process can typically include the transformation between electricity and gas by gas turbine and power to gas facility [14], and the transformation between electricity and hydrogen by water electrolysis and fuel cells [15].

Concerning the electricity consumption, flexible adjustments of electricity loads to meet heterogeneous energy demand can be made by leveraging the inherent inertia characteristics of heterogeneous energy flows. For instance, air conditioning systems in buildings can be temporarily turned off to reduce electricity loads while the indoor temperature can be maintained to not affect users' comfort due to the thermal inertia of buildings [16]. Therefore, the interactions with heterogeneous energy flows can contribute to maintaining energy balance in the low-carbon power system when renewable integration decreases.

Consequently, coordinated energy management for the low-carbon power system can be externally achieved through interactions with external heterogeneous energy flows, when managing energy conversion resources and flexible heterogeneous energy demand.

1.3 Promising Manner to Grab the Opportunities: Resource Aggregation

1.3.1 What Is a Resource Aggregation Manner

Resource aggregation entails consolidating dispatchable resources in each tranmission sub-system, distribution sub-system, and heterogeneous energy system to act as

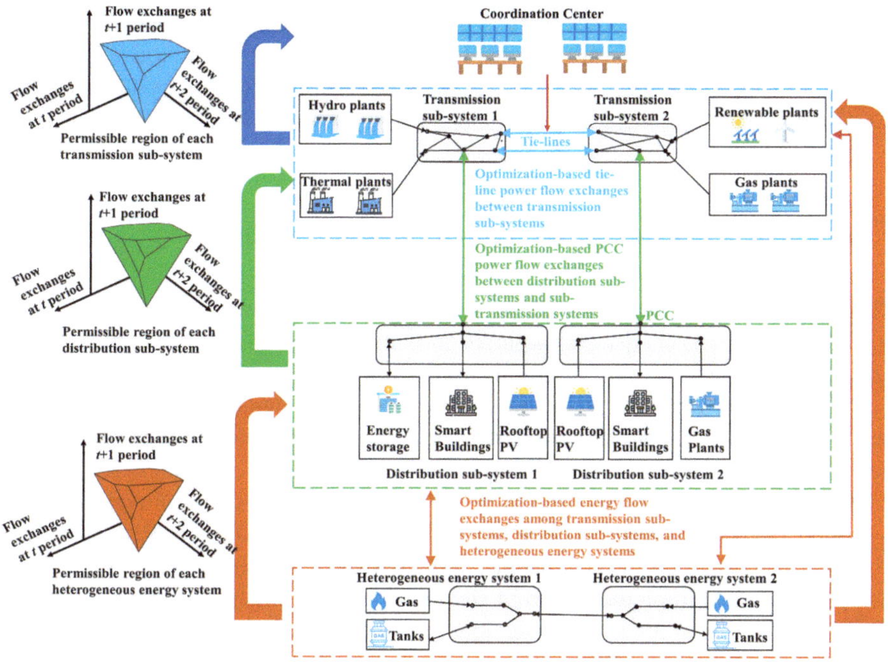

Fig. 1.5 Framework of resource aggregation-based coordinated energy management for the low-carbon power system. Note that dispatchable resources in each transmission/distribution sub-system/heterogeneous energy system include, but are not limited to, the resources in the figure

a unified entity during coordinated energy management. In this context, the framework of resource aggregation-based coordinated energy management is outlined in Fig. 1.5, comprising the following two key steps:

Step 1. Submission of aggregation information. Each transmission sub-system, distribution sub-system, and heterogeneous energy system aggregates its resources, which can be dispatched to affect the flow exchanges between itself and its interconnected counterpart(s), as permissible regions of flow exchanges. Since a low-carbon power system focuses on the electricity side, the aggregation sequence prioritizes heterogeneous energy systems, followed by distribution sub-systems, and the transmission sub-systems. Upon submitting the permissible region of tie-line power flow exchanges to the coordination center by each transmission sub-system, it comes down to Step 2 for decision-making of coordinated management.

Step 2. Decision-making of coordinated management. Decision-making on tie-line power flow exchanges is first determined by the coordination center by formulating an optimization problem subject to the permissible regions from transmission sub-systems. Given the determined tie-line power flow exchanges as boundary conditions in each transmission sub-system, the operator of each transmission sub-system then formulates an optimization problem to determine the flow exchanges to its interconnected distribution sub-systems and heterogeneous energy systems.

Given the determined flow exchanges from transmission sub-systems as boundary conditions, decision-making processes similar to those of transmission sub-systems further occur in distribution sub-systems and heterogeneous energy systems for their coordinated energy management.

1.3.2 Why to Select a Resource Aggregation Manner

This subsection first reviews other potential options in current academia and industries for coordinated energy management and subsequently distinguishes the resource aggregation manner introduced in this book from the reviewed options.

As illustrated in Sect. 1.2, coordinated energy management for the low-carbon power system encompasses transmission sub-systems, distribution sub-systems, and heterogeneous energy systems. Consequently, coordinated energy management to determine the flow exchanges among different interactions can be intuitively achieved by solving an optimization problem, which is built up by considering all the dispatchable resources, and all the parameters to formulate management constraints. However, the practical implementation of such a centralized manner [17] is challenging due to reluctance among interconnections to share parameters. This arises from their different operators; thus, either privacy worries or tedious processes to share parameters prevent their willingness. For example, in China, transmission sub-systems are operated by provincial power companies while distribution sub-systems are managed by county/municipal power companies [18]. In the United States, transmission sub-systems in different spatial regions are operated by different independent system operators [19].

As alternatives to the centralized manner, scholars have focused on distributed and decentralized manners of coordinated energy management. They decompose the centralized optimization problem into sub-problems, each of which can be independently solved by each operator. Through iterative strategies designed to adjust the flow exchanges among interconnections, coordinated energy management for the low-carbon power system can be achieved without sharing the full parameters among interactions [20].

Despite extensive studies on decentralized and distributed manners to coordinated energy management [21–23], their limited real-world applications have been observed [20]. This limitation arises from the iterative information exchanges required in decentralized and distributed manners for the flow adjustments among interconnections, while current industries prefer a non-iterative value-based manner for coordinated energy management. This value-based manner computes the maximum values of the flow exchanges among interconnections to guide coordinated energy management. For example, the *available transfer capability* (ATC), defined as the remaining amount of electric power that can be transferred on a specific path, is calculated by transmission operators to quantify potential tie-line power flow exchanges among different transmission sub-systems. These ATC values are then used to guide tie-line schedules so that transmission sub-systems can be coordinated in a non-iterative manner [24]. One more example is that FERC Order 2222 [25] facilitates DER participation in the US markets by allowing DERs in distribution sub-systems to submit their potential flow exchange values with transmission sub-systems.

Compared with the aforementioned review, the advantages of the resource aggregation manner briefly introduced in the last subsection can be highlighted below.

(1) Unlike the centralized manner, the resource aggregation manner avoids the impractical assumption of requiring the full parameters from all the transmission sub-systems, distribution sub-systems, and heterogeneous energy systems to construct a centralized optimization problem for coordinated energy management.
(2) Compared with the distributed and decentralized manners, the resource aggregation manner eliminates the need for iterative information exchanges among interconnections, making it more practical for implementation in power industries.
(3) Compared with the value-based manner in current industries, the resource aggregation manner delineates potential flow exchanges among interconnections as a permissible region instead of a single maximum value, offering more information for coordinated energy management. This characteristic provides two advantages. First, play-and-plug practical implementation is facilitated by directly displacing the value-based manner. Second, the effectiveness of the resource aggregation manner can be much better than that of the value-based manner, as numerically demonstrated in Chap. 6 by various application scenarios in the operations and planning of low-carbon power systems.

1.3.3 Key Technological Aspects

The key technological aspects of resource aggregation-based coordinated energy management for the low-carbon power system can be summarized from the following two aspects.

(1) Calculation of permissible regions: The first step in Fig. 1.5 focuses on determining the permissible regions of flow exchanges among interconnections. Therefore, a critical technological aspect is devising methods to accurately compute these permissible regions. Moreover, striking a balance between computational efficiency and accuracy is essential to maintain the effectiveness of permissible regions in operational settings. These issues will be discussed in Chaps. 2–4, which focus on designing resource aggregation techniques tailored to transmission sub-systems, distribution sub-systems, and heterogeneous energy systems. Chap. 5 further considers inevitable renewable uncertainties in resource aggregation for comparison and discussions.

(2) Utilization of permissible regions. The second step in Fig. 1.5 entails employing permissible regions for coordinated energy management in the low-carbon power system. Consequently, integrating permissible regions into coordinated energy management represents another key technological aspect. This issue will be extensively discussed in Chap. 6, which demonstrates various application scenarios in the operational and planning phases of the low-carbon power system.

1.4 Summary

In response to environmental issues posed by fossil fuel combustion, there has been a global surge in renewable installations. The resultant volatility and uncertainties of renewable integration necessitate coordinated energy management to maintain energy balance in the operations and planning of the low-carbon power system. Various manners for implementing coordinated energy management include centralized, distributed, decentralized, value-based, and resource aggregation manners. Among these manners, the resource aggregation manner stands out for its privacy protection, play-and-plug applicability in current industries, and superior coordination effectiveness. These advantages motivate the introduction of coordinated energy management for the low-carbon power system from a resource aggregation perspective in this book.

As the introductory chapter of this book, this chapter focuses on outlining coordination opportunities of coordinated energy management among transmission subsystems and distribution subsystems, and their interactions with heterogeneous energy flows. In addition, the basic framework of resource aggregation-based coordinated energy management involves computing permissible regions of flow exchanges and incorporating them into the operations and planning of the low-carbon power system. These two key technological aspects are elaborated upon in subsequent chapters of this book.

References

1. CO_2 Emissions in 2022. https://www.iea.org/reports/co2-emissions-in-2022
2. Wang, Y., Guo, C., Chen, X., et al.: Carbon peak and Carbon neutrality in China: goals, implementation path, and prospects. China Geol. **4**(4), 720–746 (2021)
3. Clean energy for America act. https://www.finance.senate.gov/download/clean-energy-for-america-one-pager
4. Germany's renewable energy sources act. https://www.iea.org/policies/12392-germanys-renewables-energy-act
5. Global installed renewable energy capacity by technology. https://ourworldindata.org/grapher/installed-global-renewable-energy-capacity-by-technology
6. Demand-side flexibility for power sector transformation. https://www.irena.org/publications/2019/Dec/Demand-side-flexibility-for-power-sector-transformation

7. Gielen, D., Boshell, F., Saygin, D., et al.: The role of renewable energy in the global energy transformation. Energy Strat. Rev. **24**, 38–50 (2019)
8. Wang, J., Chen, L., Tan, Z., et al.: Inherent spatiotemporal uncertainty of renewable power in China. Nat. Commun. **14**, 5379 (2023)
9. Zhuo, Z., Du, E., Zhang, N., et al.: Cost increase in the electricity supply to achieve Carbon neutrality in China. Nat. Commun. **13**, 3172 (2022)
10. Pehl, M., Arvesen, A., Humpenöder, F., et al.: Understanding future emissions from low-carbon power systems by integration of life-cycle assessment and integrated energy modelling. Nat. Energy **2**, 939–945 (2017)
11. Jia, Z., Wen, S., Wang, Y.: Power coming from the sky: economic benefits of inter-regional power transmission in China. Energy Econ. **119**, 106544 (2023)
12. Wang, Y.: Research on development and operation of renewable energy generation in German. Electr. Power **53**(5), 112–120 (2020)
13. Unlocking the potential of distributed energy resources: power system opportunities and best practices. https://www.iea.org/reports/unlocking-the-potential-of-distributed-energy-resources
14. Zhang, S., Gu, W., Zhang, X., et al.: Dynamic modeling and simulation of integrated electricity and gas systems. IEEE Trans. Smart Grid **14**(2), 1011–1026 (2023)
15. Hossain, M., Islam, A.M., Hossain, M.A., Hossain, M.J.: Control strategy review for hydrogen-renewable energy power system. IEEE Trans. Smart Grid **72**(15), 108170 (2023)
16. Zhang, S., Gu, W., Zhang, X., et al.: Steady-state security region of integrated energy system considering thermal dynamics. IEEE Trans. Power Syst. **39**(2), 4138–4153 (2024)
17. Tostado-Véliz, M., Liang, Y., Hasanien, H.M., et al.: Robust optimal coordination of active distribution networks and energy communities with high penetration of renewables. Renew. Energy **218**, 119286 (2023)
18. In brief: electricity distribution and sales in China. https://www.lexology.com/library/detail.aspx?g=89f7a4f8-5270-42be-9b50-80e99a3ae0d9
19. Greer, M.: Electricity marginal cost pricing: applications in eliciting demand response. Butterworth-Heinemann (2012)
20. Zuo, K., Wu, L.: A review of decentralized and distributed control approaches for islanded microgrids: novel designs, current trends, and emerging challenges. Electr. J. **35**(5), 107138 (2022)
21. Wu, J., Luh, P.B., Chen, Y., et al.: A novel optimization approach for sub-hourly unit commitment with large numbers of units and virtual transactions. IEEE Trans. Power Syst. **37**(5), 3716–3725 (2022)
22. Zhao, F., Litvinov, E., Zheng, T.: A marginal equivalent decomposition method and its application to multi-area optimal power flow problems. IEEE Trans. Power Syst. **29**(1), 53–61 (2014)
23. Kim, B.H., Baldick, R.: A comparison of distributed optimal power flow algorithms. IEEE Trans. Power Syst. **15**(2), 599–604 (2000)
24. Li, W., Vaahedi, E., Lin, Z.: BC Hydro's transmission reliability margin assessment in total transfer capability calculations. IEEE Trans. Power Syst. **28**(4), 4796–4802 (2013)
25. Participation of distributed energy resource aggregations in markets operated by regional transmission organizations and independent system operators. https://www.ferc.gov/sites/default/files/2020-09/E-1_0.pdf

Chapter 2
Resource Aggregation for Low-Carbon Power Systems at the Transmission Level

2.1 Introduction

The role of one transmission sub-system in the low-carbon power system is for renewable electricity delivery from the generation side to the load side. With the fast development of global renewable installations, energy balance in one single transmission sub-system becomes more difficult due to the inevitable power fluctuations from renewable integration [1]. This further necessitates interconnected transmission sub-systems by which rich renewable integration in one transmission sub-system can be delivered to another transmission sub-system with poor renewable integration, contributing to smoothening renewable fluctuations to simultaneously maintain energy balance [2]. Consequently, tie-line power flow exchanges among different transmission sub-systems play a vital role in the insurance of their energy balance in a broad resource configuration. In this context, the permissible region of tie-line power flow exchanges of each transmission sub-system is necessary for such coordinated energy management among different transmission sub-systems so that tie-line power flow exchanges can be determined. One permissible region of tie-line power flow exchanges of one transmission sub-system indicates that all the resources in the transmission sub-system act as a unified entity adhering to its whole constraints (e.g., power flow equations, and thermal ratings). If the permissible region of one transmission sub-system can be accurately identified—i.e., resource aggregation at the transmission level can be accurately implemented-, the transmission-level coordinated energy management can be effectively achieved. This motivates this chapter to focus on the resource aggregation methods for one transmission sub-system, i.e., identifying its permissible region of tie-line power flow exchanges.

The remaining part of this chapter is organized below. By leveraging the linearity of transmission-level constraints widely used in power industries, Sect. 2.2 will exhibit the desirable employment of the multi-parametric programming theory for an accurate identification of the permissible region of tie-line power flow exchanges. However, its computational efficiency may face various challenges from various

W. Lin et al., *Coordinated Energy Management for Low-Carbon Power Systems*,
https://doi.org/10.1007/978-981-96-7533-3_2

issues, such as the large scale of one transmission sub-systems, multiple border buses, and multiple time periods. To alleviate their impacts on the computation of identifying a permissible region of tie-line power flow exchanges, these issues will be systematically resolved based on their spatial and temporal features. Particularly, Sect. 2.3 will focus on the resource aggregation methods for a relaxed computational burden under spatial issues (including a large-scale sub-system and multiple border buses). As for the temporal issues arising from multiple time periods, Sect. 2.4 will introduce the temporal acceleration-oriented resource aggregation methods for a relaxed computational burden under multiple time periods. Finally, the summary of this chapter will be provided in Sect. 2.5.

2.2 Multi-parametric Programming-Based Resource Aggregation

2.2.1 *Problem Formulation*

Resource aggregation at the transmission level comes from an *economic dispatch* (ED) model. In accordance with the current industrial practice, the DC power flow model [3, 4] is used to construct the following ED model:

(1) **Objective function of one ED model**

The objective function of one ED model generally seeks to minimize the operational cost, i.e.,

$$\min_{\mathbf{P}_{\mathrm{G}},\theta} z = f\left(\mathbf{P}_{\mathrm{G}}\right), \tag{2.1}$$

where $\mathbf{P}_{\mathrm{G}}$ stacks active generation levels; $\boldsymbol{\theta}$ is the phase angles; f is the cost function, which is generally a linear cost function [5] or a quadratic cost function [6]. Note that: (1) Section 2.2 takes a one-segment linear cost function for illustration purposes; (2) $\boldsymbol{\theta}$ can be further categorized into $\boldsymbol{\theta}_{\mathrm{E}}$ associated with internal buses of the focused transmission sub-system) and $\boldsymbol{\theta}_{\mathrm{B}}$ associated with border buses.

(2) **Transmission-level constraints of one ED model**

a. Power balance at border buses

$$\mathbf{M}_{\text{B-LineB}}\mathbf{P}_{\mathrm{B}} = \mathbf{M}_{\text{B-LineEB}}\left(\mathbf{B}_{\text{LineEBE}}\boldsymbol{\theta}_{\mathrm{E}} + \mathbf{B}_{\text{LineEBB}}\boldsymbol{\theta}_{\mathrm{B}}\right) - \mathbf{P}_{\mathrm{DB}}, \tag{2.2}$$

where $\mathbf{P}_{\mathrm{B}}$ is tie-line power flow exchanges; $\boldsymbol{\theta}_{B}$ is phase angles of the border buses; $\mathbf{P}_{\mathrm{DB}}$ is power demands at border buses; $\mathbf{M}_{\text{B-LineB}}$ is the node-branch connection matrix between border buses and their connected lines; $\mathbf{M}_{\text{B-LineEB}}$ is the node-branch connection matrix between border buses and the lines disconnected to border buses; $\mathbf{B}_{\text{LineEBE}}$ is the branch admittance matrix between internal buses and lines connected to border buses; $\mathbf{B}_{\text{LineEBB}}$ is the branch admittance matrix between border buses and their connected lines.

b. Power balance at internal buses

$$\mathbf{T}_{\text{EG}}\mathbf{P}_{\text{G}} - \mathbf{P}_{\text{DE}} = \mathbf{B}_{\text{EE}}\boldsymbol{\theta}_{\text{E}} + \mathbf{B}_{\text{EB}}\boldsymbol{\theta}_{\text{E}}, \tag{2.3}$$

where $\mathbf{P}_{\text{DE}}$ is power demands at internal buses; $\mathbf{T}_{\text{EG}}$ is the connection matrix between border buses and generators; $\mathbf{B}_{\text{EE}}$ is the node admittance matrix among internal buses; $\mathbf{B}_{\text{EB}}$ is the node admittance matrix between internal buses and border buses.

c. Generator capacities

$$\underline{\mathbf{P}}_{\text{G}} \leq \mathbf{P}_{\text{G}} \leq \overline{\mathbf{P}}_{\text{G}}, \tag{2.4}$$

where $\underline{\mathbf{P}}_{\text{G}}$ and $\overline{\mathbf{P}}_{\text{G}}$ are the lower and upper bounds of active generation levels.

d. Tie-line power flow exchange capacities

$$\underline{\mathbf{P}}_{\text{B}} \leq \mathbf{P}_{\text{B}} \leq \overline{\mathbf{P}}_{\text{B}}, \tag{2.5}$$

where $\underline{\mathbf{P}}_{\text{B}}$ and $\overline{\mathbf{P}}_{\text{B}}$ are the thermal lower and upper bounds of tie-line power flow exchanges.

e. Phase angle limitations

$$-\pi \leq \boldsymbol{\theta} \leq \pi, \tag{2.6}$$

f. Branch flow limitations

$$\underline{\mathbf{P}}_{\text{LineEB}} \leq \mathbf{B}_{\text{LineEBE}}\,\boldsymbol{\theta}_{\text{E}} + \mathbf{B}_{\text{LineEBB}}\,\boldsymbol{\theta}_{\text{B}} \leq \overline{\mathbf{P}}_{\text{LineEB}}\,, \tag{2.7}$$

$$\underline{\mathbf{P}}_{\text{LineEE}} \leq \mathbf{B}_{\text{LineEEE}}\,\boldsymbol{\theta}_{\text{E}} + \mathbf{B}_{\text{LineEEB}}\boldsymbol{\theta}_{\text{B}} \leq \overline{\mathbf{P}}_{\text{LineEE}}\,, \tag{2.8}$$

where $\underline{\mathbf{P}}_{\text{LineEB}}$ and $\overline{\mathbf{P}}_{\text{LineEB}}$ are the lower and upper bounds of power flows of the branches connected to border buses; $\mathbf{P}_{\text{LineEE}}^{\text{min}}$ and $\mathbf{P}_{\text{LineEE}}^{\text{max}}$ are the lower and upper bounds of power flows of the branches disconnected to border buses; $\mathbf{B}_{\text{LineEEE}}$ is the branch admittance matrix among internal buses; $\mathbf{B}_{\text{LineEEB}}$ is the branch admittance matrix between internal buses and border buses.

For convenience of discussions, the ED model (2.1)–(2.8) is compacted as the following unified formulation:

$$\min_{\mathbf{x}} z = \mathbf{C}^{\top}\mathbf{x}, \tag{2.9}$$

$$\text{s. t. } \mathbf{A}\mathbf{x} \leq \mathbf{B}\mathbf{w} + \mathbf{D}, \tag{2.10}$$

where $\mathbf{x}$ is decision variables; $\mathbf{w}$ is coupling parameters; the matrices $\mathbf{A}$ and $\mathbf{B}$ and the vector $\mathbf{D}$ are constants to form the constraints (2.10); the vector $\mathbf{C}$ in the objective function (2.9) stacks the cost coefficients for decision variables $\mathbf{x}$. Note that (1) the coupling parameters may include tie-line power flow exchanges $\mathbf{P}_{\mathrm{B}}$ and border phase angles $\boldsymbol{\theta}_{\mathrm{B}}$ based on different types of tie-lines[1] [4] while they are collectively referred to as permissible regions of tie-line power flow exchanges in this book, and (2) the decisions variables $\mathbf{x}$ include all the variables in (2.1)–(2.8) except coupling parameters.

For the transmission-level resource aggregation, its task is to identify one permissible region over the coupling parameters $\mathbf{w}$ with which the transmission-level constraints (2.10) can be fulfilled. Section 2.2.2 will further demonstrate how to accomplish this task by leveraging the multi-parametric programming theory in [7].

2.2.2 Aggregation Accomplishment Based on Multi-parametric Programming

Multi-parametric programming-based resource aggregation for one transmission sub-system starts from the following definitions and properties provided in [7].

Definition 2.1 (*Optimal Partition*) Denote the optimal solutions of the optimization problem (2.9) and (2.10) as $\mathbf{x}^*$. The active and inactive constraints in the optimization problem (2.9) and (2.10) can be expressed as follows:

$$\mathbf{A}_{\mathrm{A}}\mathbf{x}^* = \mathbf{B}_{\mathrm{A}}\mathbf{w} + \mathbf{D}_{\mathrm{A}} \tag{2.11}$$

$$\mathbf{A}_{\mathrm{I}}\mathbf{x}^* < \mathbf{B}_{\mathrm{I}}\mathbf{w} + \mathbf{D}_{\mathrm{I}}, \tag{2.12}$$

where the subscripts A and I indicate the constraint indexes associated with the active and inactive constraints, respectively. Note that any constraint in (2.10) should belong to either active constraints or inactive constraints, i.e., $\mathbf{A} = \left[\, \mathbf{A}_{\mathrm{A}}^\top \ \mathbf{A}_{\mathrm{I}}^\top \,\right]^\top$, $\mathbf{B} = \left[\, \mathbf{B}_{\mathrm{A}}^\top \ \mathbf{B}_{\mathrm{I}}^\top \,\right]^\top$, and $\mathbf{D} = \left[\, \mathbf{D}_{\mathrm{A}}^\top \ \mathbf{D}_{\mathrm{I}}^\top \,\right]^\top$. ■

Definition 2.2 (*Critical Region*) Consider the following *Karush-Kuhn-Tucker* (KKT) conditions of the optimization problem (2.9) and (2.10) [8]:

$$\mathbf{C} + \mathbf{A}^\top \boldsymbol{\lambda} = \mathbf{0}, \tag{2.13}$$

$$\boldsymbol{\lambda} \otimes (\mathbf{A}\mathbf{x} - \mathbf{B}\mathbf{w} - \mathbf{D}) = \mathbf{0}, \tag{2.14}$$

[1] This book may demonstrate different resource aggregation methods at the transmission level using different types of tie-lines without being specified in advance.

$$\mathbf{Ax} - \mathbf{Bw} - \mathbf{D} \leq \mathbf{0}, \tag{2.15}$$

$$\boldsymbol{\lambda} \geq \mathbf{0}, \tag{2.16}$$

where $\boldsymbol{\lambda}$ is the dual variables of the constraints (2.10); the operator $\otimes$ is the Hardmard product. Particularly, the constraints (2.13) are called the stationary conditions in the KKT conditions; the constraints (2.14) are called the complementary conditions in the KKT conditions; the constraints (2.15) are called the primal feasibility conditions in the KKT conditions; the constraints (2.16) are called the dual feasibility conditions in the KKT conditions.

Once the coupling parameters $\mathbf{w}$ are given as $\mathbf{w}_i$, the optimal solutions $\mathbf{x}^*$ of the optimization problem (2.9) and (2.10) and their resultant active constraints (2.11) and inactive constraints (2.12) can be identified. In this context, the KKT conditions (2.13)–(2.16) can be reformulated as

$$\mathbf{C} + \mathbf{A}_{\mathrm{A}}^{\top}\boldsymbol{\lambda}_{\mathrm{A}} + \mathbf{A}_{\mathrm{I}}^{\top}\boldsymbol{\lambda}_{\mathrm{I}} = \mathbf{0}, \tag{2.17}$$

$$\mathbf{A}_{\mathrm{A}}\mathbf{x}^* = \mathbf{B}_{\mathrm{A}}\mathbf{w} + \mathbf{D}_{\mathrm{A}}, \tag{2.18}$$

$$\boldsymbol{\lambda}_{\mathrm{I}} = \mathbf{0}, \tag{2.19}$$

$$\mathbf{A}_{\mathrm{I}}\mathbf{x}^* < \mathbf{B}_{\mathrm{I}}\mathbf{w} + \mathbf{D}_{\mathrm{I}}, \tag{2.20}$$

$$\boldsymbol{\lambda}_{\mathrm{A}} \geq \mathbf{0}, \tag{2.21}$$

where $\boldsymbol{\lambda}_{\mathrm{A}}$ and $\boldsymbol{\lambda}_{\mathrm{I}}$ are the dual variables associated with the active and inactive constraints, respectively.

Suppose that the optimal solutions $\mathbf{x}^*$ of the optimization problem (2.9) and (2.10) are unique. In this context, the matrix $\mathbf{A}_{\mathrm{A}}$ is a non-singular matrix, i.e., its rank number is the same as the dimension of $\mathbf{x}^*$. Consequently, solving the linear Eq. (2.18) (by regarding $\mathbf{x}^*$ as variables) can lead to the following mapping function from the coupling parameters $\mathbf{w}$ to the optimal solutions $\mathbf{x}^*$:

$$\mathbf{x}^*(\mathbf{w}) = \mathbf{A}_{\mathrm{A}}^{-1}\mathbf{B}_{\mathrm{A}}\mathbf{w} + \mathbf{A}_{\mathrm{A}}^{-1}\mathbf{D}_{\mathrm{A}} = \mathbf{G}_{\mathrm{x},i}\mathbf{w} + \mathbf{F}_{\mathrm{x},i}. \tag{2.22}$$

When degeneracy occurs in the optimization problem (2.9) and (2.10), the uniqueness of the optimal solutions $\mathbf{x}^*$ can be ensured by the consistent tie-breaking rule in [9] and then the mapping function (2.22) can also be obtained.

Let's further substitute (2.22) into (2.20) to get

$$\begin{aligned}
&\mathbf{A}_{\mathrm{I}}\mathbf{x}^*(\mathbf{w}) < \mathbf{B}_{\mathrm{I}}\mathbf{w} + \mathbf{D}_{\mathrm{I}},\\
&\Rightarrow \mathbf{A}_{\mathrm{I}}\left(\mathbf{G}_{\mathrm{x},i}\mathbf{w} + \mathbf{F}_{\mathrm{x},i}\right) < \mathbf{B}_{\mathrm{I}}\mathbf{w} + \mathbf{D}_{\mathrm{I}},\\
&\Rightarrow \mathbf{A}_{\mathrm{I}}\mathbf{G}_{\mathrm{x},i}\mathbf{w} + \mathbf{A}_{\mathrm{I}}\mathbf{F}_{\mathrm{x},i} < \mathbf{B}_{\mathrm{I}}\mathbf{w} + \mathbf{D}_{\mathrm{I}},\\
&\Rightarrow \left(\mathbf{A}_{\mathrm{I}}\mathbf{G}_{\mathrm{x},i} - \mathbf{B}_{\mathrm{I}}\right)\mathbf{w} < \mathbf{D}_{\mathrm{I}} - \mathbf{A}_{\mathrm{I}}\mathbf{F}_{\mathrm{x},i},\\
&\Rightarrow \mathbf{G}_{\mathrm{w},i}\mathbf{w} < \mathbf{F}_{\mathrm{w},i}.
\end{aligned} \tag{2.23}$$

When the coupling parameters $\mathbf{w}$ satisfy (2.23), the KKT conditions (2.17)–(2.21) hold. Since (2.23) is indeed a polytope represented by a series of linear constraints, (2.23) is mathematically called a critical region. For each critical region, the mapping function in (2.22) can be obtained. ■

Property 2.1 *Denote the* ith *critical region from the optimization problem (2.9) and (2.10) as* $\mathbf{W}_i \triangleq \left\{\mathbf{w} \mid \mathbf{G}_{w,i}\mathbf{w} \leq \mathbf{F}_{w,i}\right\}$. *Any two critical regions are mutually exclusive. Moreover, the union of all the critical regions, i.e.,* $\mathbf{W}_F = \mathbf{W}_1 \cup \mathbf{W}_2 \cup \cdots \cup \mathbf{W}_N$, *can be represented as the following polytope:*

$$\mathbf{W}_F \triangleq \{\mathbf{w} \mid \mathbf{G}_w\mathbf{w} \leq \mathbf{F}_w\}, \tag{2.24}$$

the constant matrix $\mathbf{G}_w$ *and the constant vector* $\mathbf{F}_w$ *can be easily obtained after computing all the critical regions.* ■

Based on Definitions 2.1 and 2.2, it is not hard to know that when the coupling parameters $\mathbf{w}$ vary within a compact set $\mathbf{W}$, the set $\mathbf{W}$ can be partitioned into several critical regions, each of which is the sub-region of the permissible region of tie-line power flow exchanges. Based on Property 1, all the critical regions can be further unioned as the convex polytope (2.24). Such a convex polytope actually serves as the permissible region of tie-line power flow exchanges of the focused transmission sub-system.

Remark 2.1 Substituting (2.22) into the objective function (2.10) leads to the following mapping function from the coupling parameters $\mathbf{w}$ to the optimal objective function z^*:

$$\begin{aligned}
z^*(\mathbf{w}) &= \mathbf{C}^\top\mathbf{x}^*(\mathbf{w})\\
&= \mathbf{C}^\top\left(\mathbf{A}_{\mathrm{A}}^{-1}\mathbf{B}_{\mathrm{A}}\mathbf{w} + \mathbf{A}_{\mathrm{A}}^{-1}\mathbf{D}_{\mathrm{A}}\right)\\
&= \mathbf{C}^\top\mathbf{A}_{\mathrm{A}}^{-1}\mathbf{B}_{\mathrm{A}}\mathbf{w} + \mathbf{C}^\top\mathbf{A}_{\mathrm{A}}^{-1}\mathbf{D}_{\mathrm{A}}\\
&= \mathbf{G}_{\mathrm{z},i}\mathbf{w} + \mathbf{F}_{\mathrm{z},i}.
\end{aligned} \tag{2.25}$$

Moreover, the union of all the functions (2.25) over the critical regions is convex [7]. Such a feature eases the preservation of the cost information over the permissible region of tie-line power flow exchanges, facilitating the subsequent cost-oriented applications in Chap. 6. ■

Based on the principles above, the basic framework of multi-parametric programming-based resource aggregation for one transmission sub-system is illustrated in Fig. 2.1, and its relevant basic algorithm is given in Fig. 2.2.

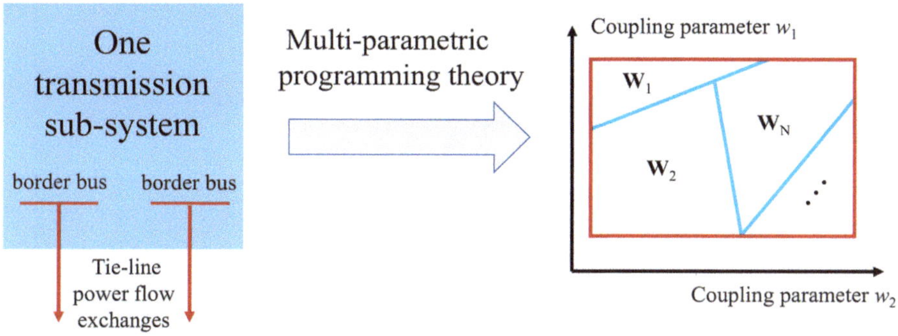

Fig. 2.1 Basic framework of multi-parametric programming-based resource aggregation for one transmission sub-system, reprinted from [4], with permission from Elsevier

2.2.3 *Case Studies*

2.2.3.1 Basic Information

The resource aggregation method in Sect. 2.2.2 will be exhibited in the IEEE 14-bus and IEEE 118-bus test systems. The system parameters have been uploaded to [10]. Particularly, the IEEE 14-bus test system is divided into two sub-systems via three tie-lines from bus 9 to bus 4, from bus 6 to bus 5, and from bus 9–7. Sub-system 1 in the IEEE 14-bus test system contains buses 1–8 while Sub-system 2 in the IEEE 14-bus test system contains buses 10–14. The IEEE 118-bus test system is similarly divided into two sub-systems via three tie-lines from bus 81 to bus 68, from bus 82 to bus 77, and from bus 84–75. Sub-system 1 in the IEEE 118-bus test system contains buses 80, 83, and 85–112, while Sub-system 2 in the IEEE 118-bus test system contains buses 1–79 and buses 113–118.

In addition to the permissible region of tie-line power flow exchanges from the resource aggregation method in Sect. 2.2.2, The value-based description commonly in current power industries for the permissible region of tie-line power flow exchanges will also be obtained for comparison. Particularly, such a value-based description will be obtained by the method in [11] which computes the maximum tie-line power flow exchange value of the border interface and each tie-line.

2.2.3.2 Comparison of Permissible Regions of Tie-Line Power Flow Exchanges Under Different Methods

Sub-systems 1 in the IEEE 14-bus and IEEE 118-bus test systems are taken as examples for the comparison of permissible regions of tie-line power flow exchanges under different methods. For Sub-system 1 in the IEEE 14-bus test system, Fig. 2.3 demonstrates the 3-D permissible region of the three tie-line power flow exchanges P_{6-5}, P_{9-4}, P_{9-7} where P_{i-j} indicates the tie-line power flow exchanges from border bus

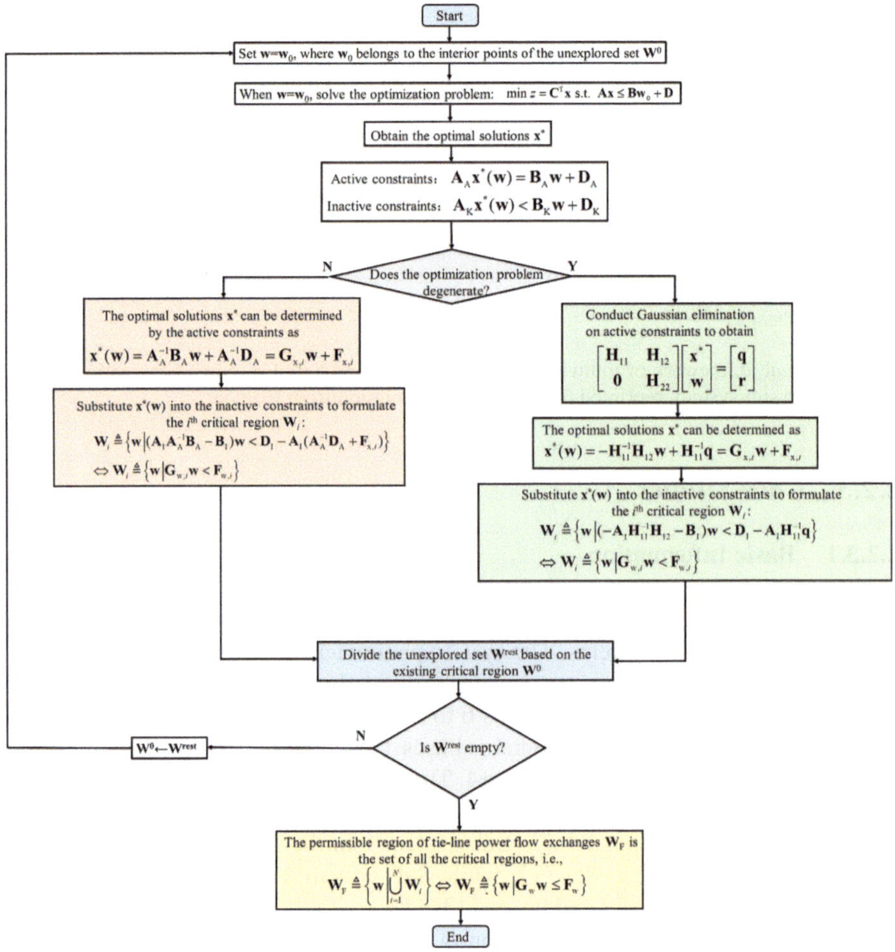

Fig. 2.2 Basic algorithm for multi-parametric programming-based resource aggregation, reprinted from [4], with permission from Elsevier

i to border bus j. Figure 2.4 demonstrates the 3-D permissible region of the three tie-line power flow exchanges P_{82-77}, P_{81-68}, P_{84-74} for Sub-system 1 in the IEEE 118-bus test system.

In Fig. 2.3, the exactness of the permissible region from the resource aggregation method in Sect. 2.2.2 will be shortly explained in Sect. 2.2.3.3. The permissible region from the value-based description method in [11] is constructed by four specific planes (i.e., the maximum value plane of P_{6-5}, the maximum value plane of P_{9-4}, the maximum value plane of P_{9-7}, and the maximum value plane of $P_{6-5} + P_{9-4} + P_{9-7}$) which are optimized subject to the transmission-level constraints (2.10). Since these four specific planes have not fully depicted the tie-line power flow exchange limitations of other tie-line power flow exchange combinations, fewer limitations are

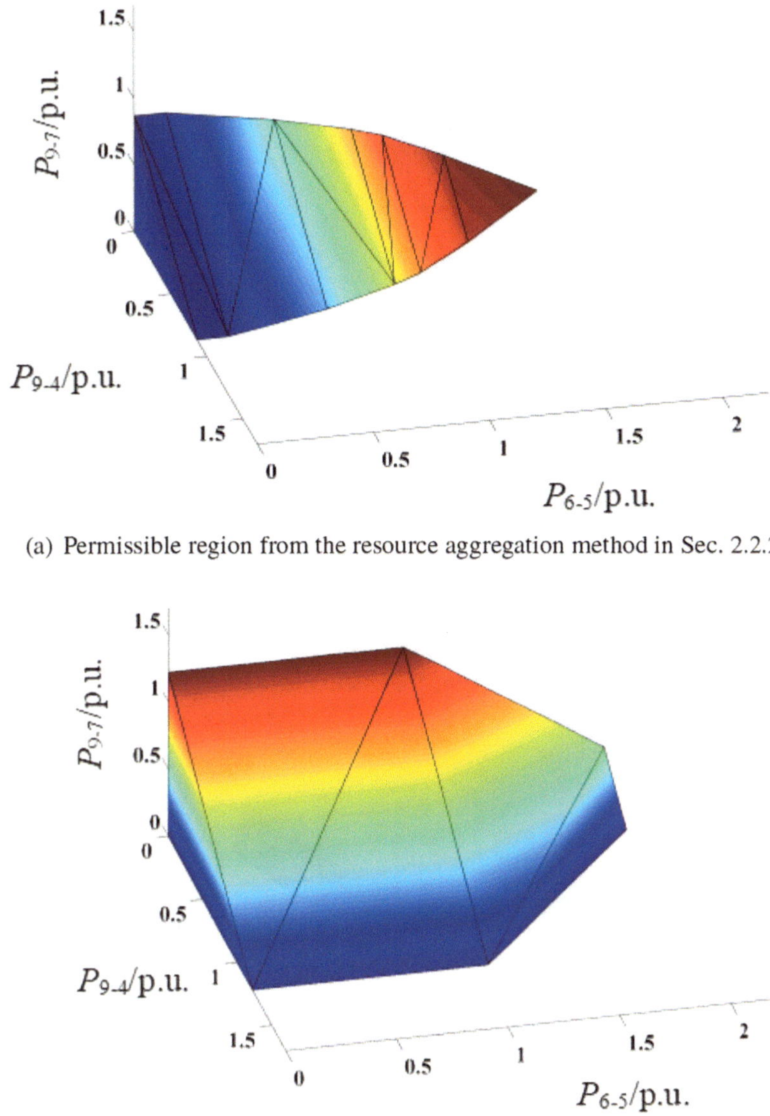

(a) Permissible region from the resource aggregation method in Sec. 2.2.2

(b) Permissible region from the value-based description method in [11]

Fig. 2.3 Permissible regions of tie-line power flow exchanges P_{6-5}, P_{9-4}, P_{9-7} in Sub-system 1 of the IEEE 14-bus test system under different methods, reprinted from [4], with permission from Elsevier

applied to the permissible region. Consequently, the permissible region from such a value-based description is larger than that of the resource aggregation method in Sect. 2.2.2 which can exactly capture the permissible region of tie-line power flow

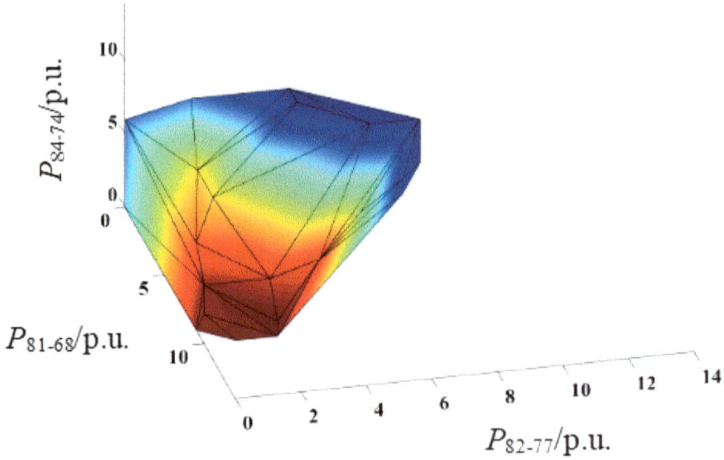

(a) Permissible region from the resource aggregation method in Sec. 2.2.2

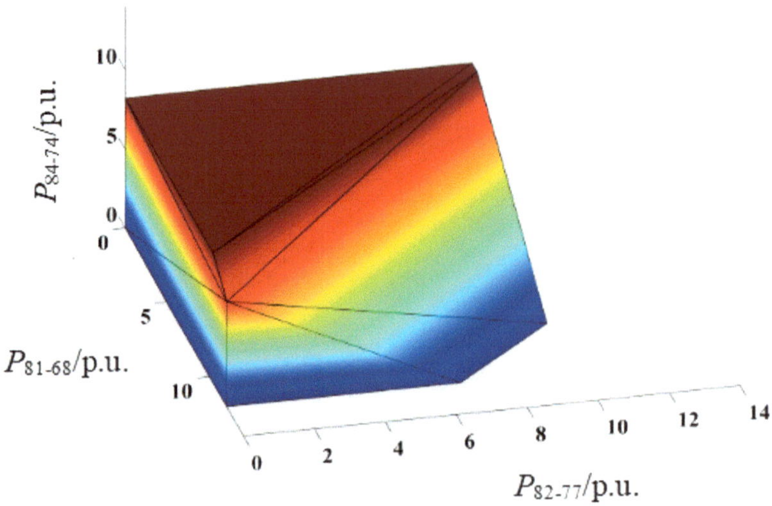

(b) Permissible region from the value-based description method in [11]

Fig. 2.4 Permissible regions of tie-line power flow exchanges P_{81-68}, P_{82-77}, P_{84-75} in Region 1 of the IEEE 118-bus test system under different methods, reprinted from [4], with permission from Elsevier

exchanges. Figure 2.4 shows the same conclusion as Fig. 2.3: the permissible region from the value-based description method in [11] is larger than the permissible region from the resource aggregation method in Sect. 2.2.2.

2.2.3.3 Effectiveness of Permissible Regions of Tie-Line Power Flow Exchanges

In the IEEE 14-bus and IEEE 118-bus test systems, let's assume that Sub-system 1 has lower generation costs and then sells electricity to Sub-system 2, where Sub-system 2's purchase cost is the selling price multiplied by the tie-line power flow exchanges from Sub-system 1. The ED model for Sub-system 2 considering external electricity purchases can be conceptually described as follows:

(1) In terms of constraints, Sub-system 2 should not only respect its transmission-level constraints (2.10) but also the permissible region of the tie-line power flow exchanges from Sub-system 1.
(2) Regarding the objective function, the total operational cost of Sub-system 2 includes the generator costs in Sub-system 2 and the electricity purchase costs from Sub-system 1.

The optimal solutions of the ED model which considers the transmission-level constraints of both Sub-system 1 and Sub-system 2 and aim to minimize the total operational cost of Sub-system 2, are taken as a benchmark. It is not hard to know: if the permissible region of the tie-line power flow exchanges in Sub-system 1 is accurately delineated, then the results of the ED model for Sub-system 2 considering external electricity purchases, after replacing the transmission-level constraints in Sub-system 1 with its permissible region, should be identical to the benchmark; otherwise, the results will deviate from the benchmark. Based on the rationales above, the optimal solutions of Sub-system 2 using different Sub-system 1 forms (i.e., either transmission-level constraints or the permissible region) under the methods in Sect. 2.2.2 and [11] are compared in Tables 2.1 and 2.2. As shown in Tables 2.1 and 2.2, the solutions of the resource aggregation method in Sect. 2.2.2 are the same as those of the benchmark, while the solutions of the method in [11] deviate from those of the benchmark. This indicates the exactness of the resource aggregation method in Sect. 2.2.2.

Table 2.1 Cost comparison in Sub-system 2 under different methods, reprinted from [18], with permission from Elsevier

Test system	Methods	Total cost ($)	Purchase cost ($)	Generation cost ($)
IEEE 14-bus	Benchmark	13349	3660	9689
	Section 2.2.2	13349	3660	9689
	Reference [11]	12365	4359	8006
IEEE 118-bus	Benchmark	85622	41092	44530
	Section 2.2.2	85622	41092	44530
	Reference [11]	83386	45849	37537

Table 2.2 Error comparison of the active generation levels and branch flows in Sub-system 2 under different methods, reprinted from [18], with permission from Elsevier

Test system	Methods	Ave. absolute error of generation levels (MW)	Ave. absolute error of branch flows (MW)
IEEE 14-bus	Section 2.2.2	0	0
	Reference [11]	68	37
IEEE 118-bus	Section 2.2.2	0	0
	Reference [11]	42	41

2.3 Spatial Acceleration-Oriented Resource Aggregation

Traditional multi-parametric programming in Sect. 2.2 needs to enumerate all the possible combinations of active and inactive constraints of the ED model in a transmission sub-system for its resource aggregation. However, the tremendous constraints in large-scale transmission sub-systems may exceptionally increase the combination number of active and inactive constraints, resulting in a heavy computation for transmission-level resource aggregation within a limited operational time. Consequently, Sect. 2.3.1 will introduce a modified multi-parametric programming algorithm to accelerate resource aggregation in large-scale transmission sub-systems.

In addition, multiple border buses in one transmission sub-system indicate a high dimension of the permissible region of tie-line power flow exchanges. Generally, the description of a high-dimension polytope requires tremendous inequalities [12], resulting in a heavy computational burden within a limited operational time. In this context, Sect. 2.3.2 will discuss an acceleration strategy to alleviate the computational burden arising from multiple border buses.

2.3.1 Acceleration in Large-Scale Transmission Sub-Systems

Let's recall the following ED model of a transmission sub-system for its resource aggregation:

$$\min_{\mathbf{x}} f(\mathbf{x}), \tag{2.26}$$

$$\text{s.t. } [\boldsymbol{\lambda}] : \mathbf{A}\mathbf{x} \leq \mathbf{B}\mathbf{w} + \mathbf{D}, \tag{2.27}$$

where $\mathbf{x}$ is decision variables; $\mathbf{w}$ is coupling parameters; $\boldsymbol{\lambda}$ is dual variables associated with the transmission-level constraints (2.27); the matrices $\mathbf{A}$ and $\mathbf{B}$ and the vector $\mathbf{D}$ are constants to form the transmission-level constraints (2.27); the cost function in (2.26) is taken as $0.5 \times \mathbf{x}^T\mathbf{H}_1\mathbf{x} + \mathbf{H}_2^T\mathbf{x}$ where $\mathbf{H}_1$ and $\mathbf{H}_2$ are the coefficients of quadratic and linear terms, respectively.

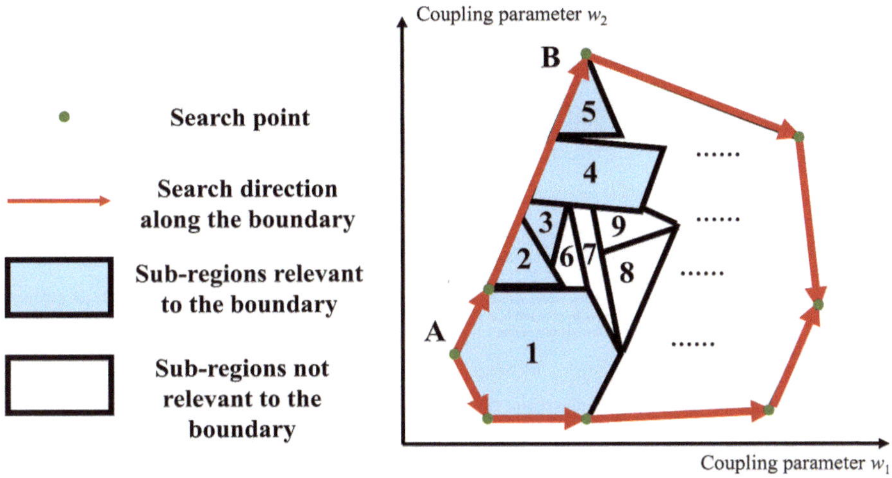

Fig. 2.5 Outline of modified multi-parametric programming-based resource aggregation, reprinted from [13], with permission from IEEE

The key issue of accelerating multi-parametric programming-based resource aggregation in a large-scale transmission sub-system is to reduce the enumeration number of the combinations of active and reactive constraints, while the permissible region of tie-line power flow exchanges can still be found. This issue can be resolved by the modified multi-parametric programming-based aggregation method in Fig. 2.5. To implement the computation in Fig. 2.5, the following three strategies are introduced to modify multi-parametric programming:

- **Strategy 1: Boundary search direction**. For traditional multi-parametric programming, the red lines (i.e., the boundary of the permissible region of tie-line power flow exchanges) in Fig. 2.5 can be obtained if and only if all the combinations of active and inactive constraints are enumerated. This strategy fixes the search direction along the boundary of the permissible region of tie-line power flow exchanges, which only enumerates the combinations of active and inactive constraints relevant to the boundary and thus avoids useless enumerations of active and inactive constraints for the inner sub-regions. For example, the inner sub-regions 6, 7, 8, and 9 will not explored by this strategy.
- **Strategy 2: Skipping abundant sub-regions**. Although the search direction can be fixed along the boundary of the permissible region of tie-line power flow exchanges, there may be a few sub-regions that share the same boundary facet. Consequently, skipping these useless sub-regions can contribute to computational acceleration. For example, the boundary facet **AB** in Fig. 2.5 is shared by the sub-regions 1, 2, 3, 4, and 5. The boundary facet **AB** can be exactly determined by only finding the sub-regions 1 and 2 when the sub-regions 3, 4, and 5 are skipped.
- **Strategy 3: Guess on KKT conditions**. In multi-parametric programming, optimization computations are required to determine sub-regions arising from KKT

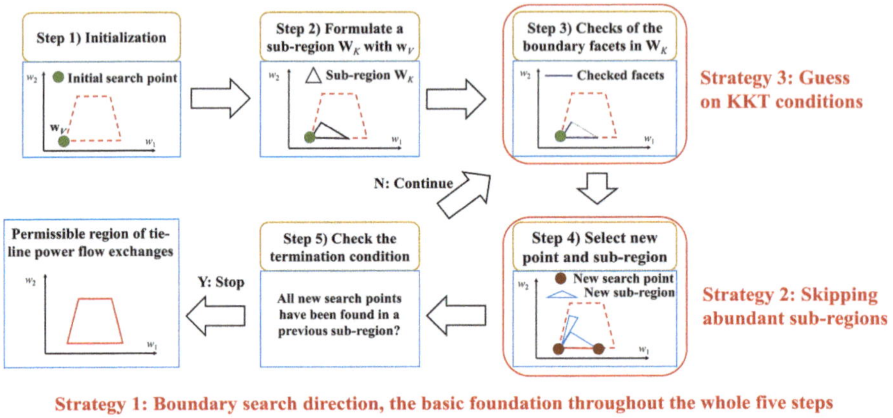

Fig. 2.6 Flowchart of modified multi-parametric programming-based resource aggregation, reprinted from [13], with permission from IEEE

conditions. For the optimization problem (2.26)–(2.27), it will be exhibited that the number of optimization computations for certain sub-regions can be avoided using a fast guess strategy for the KKT conditions. For these sub-regions, the optimization computations can be replaced by simple matrix manipulations for computational acceleration in multi-parametric programming-based resource aggregation.

Figure 2.6 details the five steps of modified multi-parametric programming-based resource aggregation especially designed for one large-scale transmission subsystem. Strategy 1 is the foundation throughout the whole process. Strategies 2 and 3 are involved in Steps 4 and 3, respectively. These five steps are clarified in the coming content.

2.3.1.1 Step 1. Initialization

Traditional multi-parametric programming-based resource aggregation starts from any interior point of the permissible region of tie-line power flow exchanges. Differently, modified multi-parametric programming-based resource aggregation starts from any vertex $\mathbf{w}_V$ of the permissible region of tie-line power flow exchanges. Its starting point can be computed using the following *linear programming* (LP) problem:

$$\min_{\mathbf{x},\mathbf{w}} \mathbf{A}_{\mathrm{w}}^{\top}\mathbf{w}, \tag{2.28}$$

$$\text{s.t. Constraints (2.27)}, \tag{2.29}$$

where $\mathbf{A}_{\mathrm{w}}$ is a vector whose dimension is the same as that of $\mathbf{w}$, and at least one element in $\mathbf{A}_w$ is non-zero.

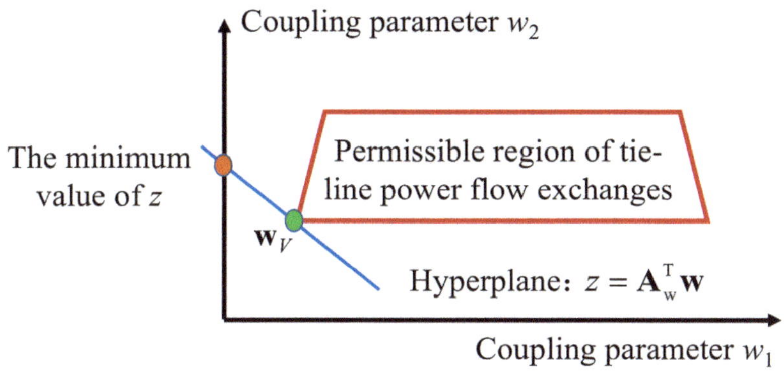

Fig. 2.7 Graphic illustration of the LP problem (2.28) and (2.29), reprinted from [13], with permission from IEEE

The graphic illustration of the LP problem (2.28) and (2.29) is illustrated in Fig. 2.7. The LP problem (2.28) and (2.29) utilizes the plane $z = \mathbf{A}_w^T \mathbf{w}$ to scan the whole permissible regions. For an LP problem, when the minimum value of z is found, the resultant unique solutions $\mathbf{w}$ can be a vertex of the LP problem (2.28) and (2.29).

2.3.1.2 Step 2. Closed-Form Expression of a Sub-Region $\mathbf{W}_K$ Containing $\mathbf{w}_V$

In multi-parametric programming, a sub-region consists of tie-line power flow exchanges with which the active and inactive constraints are the same in the optimization problem (2.26) and (2.27). Suppose that the optimal solutions of (2.26) and (2.27) are $\mathbf{x}^*$ when $\mathbf{w} = \mathbf{w}_V$. Then, the active and inactive constraints can be identified as

$$\text{Active constraint set } \mathrm{A}_K : \mathbf{A}_{\mathrm{A}_K}\mathbf{x}^* = \mathbf{B}_{\mathrm{A}_K}\mathbf{w} + \mathbf{D}_{\mathrm{A}_K}, \tag{2.30}$$

$$\text{Inactive constraint set } \mathrm{I}_K : \mathbf{A}_{\mathrm{I}_K}\mathbf{x}^* < \mathbf{B}_{\mathrm{I}_K}\mathbf{w} + \mathbf{D}_{\mathrm{I}_K}. \tag{2.31}$$

For the active and inactive constraints (2.30) and (2.31), the KKT conditions associated with the active constraint set A_K can be formulated as

$$\mathbf{M}\begin{bmatrix} \mathbf{x}^* \\ \boldsymbol{\lambda}_{\mathrm{A}_K} \end{bmatrix} = \begin{bmatrix} -\mathbf{H}_2^\top \\ \mathbf{B}_{\mathrm{A}_K}\mathbf{w} + \mathbf{D}_{\mathrm{A}_K} \end{bmatrix}, \tag{2.32}$$

where $\mathbf{M} = \begin{bmatrix} \mathbf{H}_1 & \mathbf{A}_{\mathrm{A}_K}^\top \\ \mathbf{A}_{\mathrm{A}_K}^\top & \mathbf{0} \end{bmatrix}$; $\boldsymbol{\lambda}_{\mathrm{A}_K}$ is the dual variables associated with active constraints.

When $\mathbf{M}$ is non-singular, $\mathbf{x}^*$ and $\boldsymbol{\lambda}_{\mathrm{A}_K}$ can be derived as linear functions of $\mathbf{w}$ from (2.32), i.e.,

$$\begin{bmatrix} \mathbf{x}^*(\mathbf{w}) \\ \boldsymbol{\lambda}_{\mathrm{A}_K}(\mathbf{w}) \end{bmatrix} = \mathbf{M}^{-1} \begin{bmatrix} -\mathbf{H}_2^\top \\ \mathbf{B}_{\mathrm{A}_K}\mathbf{w} + \mathbf{D}_{\mathrm{A}_K} \end{bmatrix} = \begin{bmatrix} \mathbf{G}_\mathrm{P}\mathbf{w} + \mathbf{F}_\mathrm{P} \\ \mathbf{G}_\lambda \mathbf{w} + \mathbf{F}_\lambda \end{bmatrix}, \tag{2.33}$$

where $\mathbf{M}^{-1} = \begin{bmatrix} \mathbf{M}_{11} & \mathbf{M}_{12} \\ \mathbf{M}_{21} & \mathbf{M}_{22} \end{bmatrix}$; $\begin{bmatrix} \boldsymbol{G}_\mathrm{P} \\ \boldsymbol{G}_\lambda \end{bmatrix} = \begin{bmatrix} \mathbf{M}_{12}\mathbf{B}_{\mathrm{A}_K} \\ \mathbf{M}_{22}\mathbf{B}_{\mathrm{A}_K} \end{bmatrix}$; $\begin{bmatrix} \boldsymbol{F}_\mathrm{P} \\ \boldsymbol{F}_\lambda \end{bmatrix} = \begin{bmatrix} -\mathbf{M}_{11}\mathbf{H}_2^\top + \mathbf{M}_{12}\mathbf{D}_{\mathrm{A}_K} \\ -\mathbf{M}_{21}\mathbf{H}_2^\top + \mathbf{M}_{22}\mathbf{D}_{\mathrm{A}_K} \end{bmatrix}$.

To hold the same inactive constraints (2.31) and keep $\boldsymbol{\lambda}_{\mathrm{A}_K}(\mathbf{w}) \geq \mathbf{0}$, a sub-region $\mathbf{W}_K$ containing $\mathbf{w}_V$ is defined as

$$\mathbf{W}_K \triangleq \left\{ \mathbf{w} \mid \mathbf{A}_{\mathrm{I}_K}\mathbf{x}^*(\mathbf{w}) \leq \mathbf{B}_{\mathrm{I}_K}\mathbf{w} + \mathbf{D}_{\mathrm{I}_K}, \boldsymbol{\lambda}_{\mathrm{A}_K}(\mathbf{w}) \geq \mathbf{0} \right\}. \tag{2.34}$$

Substituting (2.33) into (2.34) leads to

$$\mathbf{W}_K \triangleq \{ \mathbf{w} \mid \boldsymbol{\alpha}\mathbf{w} + \boldsymbol{\beta} \leq \mathbf{0} \}, \tag{2.35}$$

where $\boldsymbol{\alpha} = \left[\left(\mathbf{A}_{\mathrm{I}_K}\mathbf{G}_\mathrm{P} - \mathbf{B}_{\mathrm{I}_K} \right)^\top \; -\mathbf{G}_\lambda^\top \right]^\top$; $\boldsymbol{\beta} = \left[\left(\mathbf{A}_{\mathrm{I}_K}\mathbf{F}_\mathrm{P} - \mathbf{D}_{\mathrm{I}_K} \right)^\top \; -\mathbf{F}_\lambda^\top \right]^\top$.

Note that the matrix $\mathbf{M}$ in (2.32) could be singular due to degeneracy. In this context, the method in [14] can be employed for the existence of (2.35).

2.3.1.3 Step 3. Checks of the Boundary Facets in $\mathbf{W}_K$

This step checks whether the facets where $\mathbf{w}_V$ lies are the boundary facets of the whole permissible region. It includes: (1) searching a point $\mathbf{w}_{V+1}$ that steps over the $\mathbf{W}_K$ via the ith checked facet, and (2) searching one new active constraint set A_{K+1} and one new inactive constraint set I_{K+1} for $\mathbf{w}_{V+1}$ and then formulate a sub-region $\mathbf{W}_{K+1}$ that contains $\mathbf{w}_{V+1}$ as illustrated in Step 2. If there are no other sub-regions $\mathbf{W}_{K+1}$ adjacent to $\mathbf{W}_K$ via the ith checked facet, the ith checked facet is a boundary facet of the whole permissible region.

In traditional multi-parametric programming, the pivot computation in [14] is typical to implement this step. However, it requires iterative optimization computations. To alleviate such optimization computations for a relaxed computational burden, a pre-processing stage will be added prior to implementing the pivot computation. This paves one promising way to reduce the computational burden in multi-parametric programming. Partially, the pre-processing and the pivot computation stages are presented below.

- **Pre-processing stage**. Its key idea is illustrated in Fig. 2.8. When a point $\mathbf{w}_K$ in $\mathbf{W}_K$ moves to $\mathbf{w}_V$ toward the checked facet, some of inactive constraints could become active, i.e., $\mathbf{A}_{\mathrm{I}_{KK}}\ \mathbf{x}^*(\mathbf{w}_K) - \mathbf{B}_{\mathrm{I}_{KK}}\mathbf{w}_K - \mathbf{D}_{\mathrm{I}_{KK}} < \mathbf{0} \rightarrow \mathbf{A}_{\mathrm{I}_{KK}}\mathbf{x}^*(\mathbf{w}_V) - \mathbf{B}_{\mathrm{I}_{KK}}\mathbf{w}_V - \mathbf{D}_{\mathrm{I}_{KK}} = \mathbf{0}$, where $\mathrm{I}_{KK} \subset \mathrm{I}_K$. In addition, some dual variables associated with active constraints could become zero, i.e., $\boldsymbol{\lambda}_{\mathrm{A}_{KK}}(\mathbf{w}_K) > \mathbf{0} \rightarrow \boldsymbol{\lambda}_{A_{KK}}(\mathbf{w}_V) = \mathbf{0}$, where

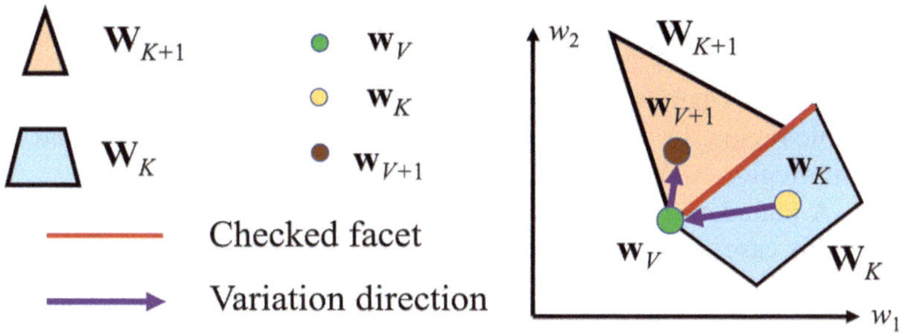

Fig. 2.8 Graphic illustration of the pre-processing stage, reprinted from [13], with permission from IEEE

$\mathrm{A}_{KK} \subset \mathrm{A}_K$. For traditional multi-parametric programming, iterative optimization computations are required to find the active constraint set A_{K+1} and inactive constraint set I_{K+1} for $\mathbf{w}_{V+1}$. In the pre-processing stage, it is assumed that the variation direction of the constraints indexed by A_{KK} and I_{KK} remains when $\mathbf{w}$ moves from $\mathbf{W}_K$ to $\mathbf{W}_{K+1}$. In this case, $(\mathrm{A}_{K+1}, \mathrm{I}_{K+1})$ are directly determined without requiring iterative optimization computations. For $\mathbf{w}_{V+1}$, its active constraint set is guessed as $\mathrm{A}_{K+1} = \mathrm{A}_K/\mathrm{A}_K K \cup \mathrm{I}_{KK}$ while its inactive constraint set is guessed as $\mathrm{I}_{K+1} = \mathrm{I}_K/\mathrm{I}_{KK} \cup \mathrm{A}_{KK}$.

- **Pivot computation stage**. This stage judges whether there is a new active constraint set A_{K+1} and an inactive constraint set I_{K+1} associated with $\mathbf{W}_{K+1}$ adjacent to $\mathbf{W}_K$ via the checked facet, which requires iterative optimization computations. If no new sets A_{K+1} and I_{K+1} can be found, the checked facet is a boundary facet of the whole permissible region; otherwise, the checked facet is not a boundary facet of the whole permissible region. The pivot computation stage ensures the judgment validity of new active and inactive constraint sets (i.e., such a combination ensures the existence of one sub-region). However, its iterative optimization computations can lead to a heavy computational burden when facing a tremendous enumeration of combinations of active and inactive constraints. Additionally, when there may be several sub-subregions that contain the checked facet, all of them will be enumerated in this pivot computation stage. However, some sub-regions are useless because they do not contain $\mathbf{w}_V$, which leads to an additional computational burden for filtering these abundant sub-regions.

2.3.1.4 Step 4. Selections of New Search Points $\mathbf{w}_{\text{next}}$ and Sub-Regions $\mathbf{W}_{\text{next}}$

This step selects $\mathbf{w}_{\text{next}}$ and $\mathbf{W}_{\text{next}}$ for the next-round search, which paves one promising way to skip the sub-regions that share the same boundary facet of the whole permissible region. Based on the results from Step 3, three different cases will be discussed to select $\mathbf{w}_{\text{next}}$ and $\mathbf{W}_{\text{next}}$.

- **Case 1**. The checked facet in $\mathbf{W}_K$ is not a boundary facet of the whole permissible region. Case 1 indicates that a new sub-set $\mathbf{W}_{K+1}$ is adjacent to $\mathbf{W}_K$ via the checked facet. The facet where $\mathbf{w}_V$ lies in $\mathbf{W}_{K+1}$ may be a boundary facet of the whole permissible region. In this case, the new search point is $\mathbf{w}_{\text{next}} = \mathbf{w}_V$ and the new search sub-region is $\mathbf{W}_{\text{next}} = \mathbf{W}_{K+1}$.
- **Case 2**. A boundary facet of the whole permissible region is found for the first time. In this case, a new search direction – the found boundary facet – has been pointed out. Consequently, other vertices along this boundary facet can serve as the next new search points, and the new search sub-region is $\mathbf{W}_{\text{next}} = \mathbf{W}_{K+1}$.
- **Case 3**. A boundary facet of the whole permissible region is found for the second time. Case 3 indicates this boundary facet is shared by a few sub-regions. In this case, skipping the potential sub-regions that share the same boundary facet can contribute to further accelerations. This can be achieved below. Vertices on the shared boundary facet will be explored along each direction $w_i (i = 1, 2, \ldots, |\mathbf{w}|$, where $|\mathbf{w}|$ is the dimension number of $\mathbf{w}$). By assuming the shared boundary facet is the ith facet in $\mathbf{W}_K$ (expressed as $\boldsymbol{\alpha}_i^{W_K}\mathbf{w} + \beta_i^{W_K} = \mathbf{0}$), the vertex exploration can be implemented by solving the following two LP problems:

$$\max_{\mathbf{x},\mathbf{w}} w_i, \text{ s.t. Constraints (2.27) and } \boldsymbol{\alpha}_i^{W_K}\mathbf{w} + \beta_i^{W_K} = \mathbf{0}, \tag{2.36}$$

$$\min_{\mathbf{x},\mathbf{w}} w_i, \text{ s.t. Constraints (2.27) and } \boldsymbol{\alpha}_i^{W_K}\mathbf{w} + \beta_i^{W_K} = \mathbf{0}. \tag{2.37}$$

The solutions of the LP problems (2.36) and (2.37) serves as the new search points $\mathbf{w}_{\text{next}}$. Then, the corresponding new search regions $\mathbf{W}_{\text{next}}$ can be formulated as shown in Step 2 when $\mathbf{w} = \mathbf{w}_{\text{next}}$.

2.3.1.5 Step 5. Termination Check

Each search point is only explored once in one sub-region. If there are no new search points, modified multi-parametric programming-based resource aggregation will terminate. This indicates that all the points have been connected by the found boundary facets to construct a closed permissible region of tie-line power flow exchanges. If there are new search points, let $\mathbf{w}_{\text{next}} \rightarrow \mathbf{w}_V$, $\mathbf{W}_{\text{next}} \rightarrow \mathbf{W}_K$ and then return to Step 3 to continue the loop.

2.3.2 *Acceleration for Multiple Border Buses*

Let's consider the following DC power flow-based constraints as an example to demonstrate the acceleration for a permissible region of tie-line power flow exchanges with multiple border buses:

$$\mathbf{e}_{\mathrm{G}}^{\top}\mathbf{P}_{\mathrm{G}} + \mathbf{e}_{\mathrm{B}}^{\top}\mathbf{P}_{\mathrm{B}} = \mathbf{e}_{\mathrm{D}}^{\top}\mathbf{P}_{\mathrm{D}}, \tag{2.38}$$

$$\mathbf{A}_{\mathrm{G}}\mathbf{P}_{\mathrm{G}} + \mathbf{A}_{\mathrm{B}}\mathbf{P}_{\mathrm{B}} + \mathbf{A}_{\mathrm{D}}\mathbf{P}_{\mathrm{D}} \leq \overline{\mathbf{P}}_{\mathrm{F}}, \tag{2.39}$$

$$\underline{\mathbf{P}}_{\mathrm{F}} \leq \mathbf{A}_{\mathrm{G}}\mathbf{P}_{\mathrm{G}} + \mathbf{A}_{\mathrm{B}}\mathbf{P}_{\mathrm{B}} + \mathbf{A}_{\mathrm{D}}\mathbf{P}_{\mathrm{D}}, \tag{2.40}$$

$$\underline{\mathbf{P}}_{\mathrm{G}} \leq \mathbf{P}_{\mathrm{G}} \leq \overline{\mathbf{P}}_{\mathrm{G}}, \tag{2.41}$$

$$\underline{\mathbf{P}}_{\mathrm{B}} \leq \mathbf{P}_{\mathrm{B}} \leq \overline{\mathbf{P}}_{\mathrm{B}}, \tag{2.42}$$

where $\mathbf{P}_{\mathrm{G}}$, $\mathbf{P}_{\mathrm{B}}$ and $\mathbf{P}_{\mathrm{D}}$ are vectors of active generation levels, tie-line power flow exchanges, and power demands, respectively; $\mathbf{e}_{\mathrm{G}}$, $\mathbf{e}_{\mathrm{B}}$ and $\mathbf{e}_{\mathrm{D}}$ are all-one vectors whose dimensions match $\mathbf{P}_{\mathrm{G}}$, $\mathbf{P}_{\mathrm{B}}$ and $\mathbf{P}_{\mathrm{D}}$, respectively; $\mathbf{Y}$ is the inverse matrix of the susceptance matrix associated with border buses; $\mathbf{M}_{\mathrm{G}}$, $\mathbf{M}_{\mathrm{B}}$, and $\mathbf{M}_{\mathrm{D}}$ are incident matrices associated with $\mathbf{P}_{\mathrm{G}}$, $\mathbf{P}_{\mathrm{B}}$ and $\mathbf{P}_{\mathrm{D}}$, respectively; $\underline{\mathbf{P}}_{\mathrm{F}}$ and $\overline{\mathbf{P}}_{\mathrm{F}}$ are the thermal lower and upper bounds of $\mathbf{P}_{\mathrm{F}}$, respectively; $\mathbf{A}_{\mathrm{G}} = \mathbf{S}\mathbf{M}_{\mathrm{G}}$; $\mathbf{A}_{\mathrm{B}} = \mathbf{S}\mathbf{M}_{\mathrm{B}}$; $\mathbf{A}_{\mathrm{D}} = \mathbf{S}\mathbf{M}_{\mathrm{D}}$; $\mathbf{S}$ is the *power transfer distribution factor* (PTDF) matrix; $\underline{\mathbf{P}}_{\mathrm{B}}$ and $\overline{\mathbf{P}}_{\mathrm{B}}$ are lower and upper bounds of $\mathbf{P}_{\mathrm{B}}$, respectively; $\underline{\mathbf{P}}_{\mathrm{G}}$ and $\overline{\mathbf{P}}_{\mathrm{G}}$ are the dispatch levels of $\mathbf{P}_{\mathrm{G}}$.

The acceleration seeks to find the permissible region of tie-line power flow exchanges feasible to the constraints (2.38)–(2.42). Particularly, to alleviate the inherent computational burden of delineating such a high-dimension permissible region due to multiple border buses, an acceleration strategy will be next introduced. Particularly, the resultant region is an inner approximation to the exact permissible region to ensure secure transmission-level operations.

The key idea of the acceleration strategy is to represent the permissible region in a new domain of $\widetilde{\mathbf{P}}_{\mathrm{B}}$ instead of the original domain of $\mathbf{P}_{\mathrm{B}}$, where the dimension of $\widetilde{\mathbf{P}}_{\mathrm{B}}$ is smaller than that of $\mathbf{P}_{\mathrm{B}}$. This can be achieved by establishing a mapping function from $\mathbf{P}_{\mathrm{B}}$ to $\widetilde{\mathbf{P}}_{\mathrm{B}}$. Particularly, such a mapping function can be explicitly established by (1) geographically categorizing tie-line power flow exchanges into several groups, and (2) bundling $\mathbf{P}_{\mathrm{B}}$ as $\widetilde{\mathbf{P}}_{\mathrm{B}}$, as will be elaborated in the coming content.

Different tie-lines are located between different transmission sub-systems. Consequently, $\mathbf{P}_{\mathrm{B}}$ can be categorized based on its geographic locations, i.e., the tie-lines, all of which connect the same two transmission sub-systems, are categorized into the same group as one equivalent tie-line. The tie-line power flow exchanges in each group are bounded as one equivalent tie-line power flow exchange as described by

$$\widetilde{P}_{\mathrm{B}}^{j} = \sum_{i=1}^{n_j} P_{\mathrm{B}}^{j_i},\ j = \{1, 2, \ldots, \mathrm{J}\}, \tag{2.43}$$

where $P_{\mathrm{B}}^{j_i}$ is the tie-line power flow exchange in $\mathbf{P}_{\mathrm{B}}$ and belongs to the jth group; $\widetilde{P}_{\mathrm{B}}^{j}$ is the bundled tie-line power exchange in the jth group; n_j is the number of the bundled tie-lines in the jth group; J is the total number of border interfaces among different transmission sub-systems.

Based on (2.43), $\mathbf{P}_{\mathrm{B}}$ can be mapped to $\widetilde{\mathbf{P}}_{\mathrm{B}} = \left[\, \widetilde{P}_{\mathrm{B}}^{1} \ldots \widetilde{P}_{\mathrm{B}}^{\mathrm{J}} \,\right]^{\top}$. This contributes to representing one permissible region in the domain of $\widetilde{\mathbf{P}}_{\mathrm{B}}$ instead of $\mathbf{P}_{\mathrm{B}}$. However, the permissible region arises from the constraints (2.38)–(2.42) which are not formulated by $\widetilde{\mathbf{P}}_{\mathrm{B}}$. To resolve this issue, Proposition 2.1 will introduce how to explicitly incorporate $\widetilde{\mathbf{P}}_{\mathrm{B}}$ into the constraints (2.38)–(2.42).

Proposition 2.1 *Suppose that there is a polytope $\widetilde{\mathbf{W}}$ defined as*

$$\widetilde{\mathbf{W}} \triangleq \left\{ \widetilde{\mathbf{P}}_B \mid \exists \left(\mathbf{P}_G, \mathbf{P}_B \right) \in \mathbf{Y} \right\}, \tag{2.44}$$

where $\mathbf{Y}$ is the following operating region revised from the constraints (2.38)–(2.42):

$$\widetilde{\mathbf{P}}_B = \widetilde{\mathbf{A}}_B \mathbf{P}_B, \tag{2.45}$$

$$\mathbf{e}_G^{\top} \mathbf{P}_G + \widetilde{\mathbf{e}}_B^{\top} \widetilde{\mathbf{P}}_B = \mathbf{e}_D^{\top} \mathbf{P}_D, \tag{2.46}$$

$$\mathbf{A}_G \mathbf{P}_G + \left(\widetilde{\boldsymbol{\alpha}}_B \widetilde{\mathbf{P}}_B + \widetilde{\boldsymbol{\beta}}_B \right) + \mathbf{A}_D \mathbf{P}_D \leq \overline{\mathbf{P}}_F - \boldsymbol{\varepsilon}_B, \tag{2.47}$$

$$\underline{\mathbf{P}}_F + \boldsymbol{\varepsilon}_B \leq \mathbf{A}_G \mathbf{P}_G + \left(\widetilde{\boldsymbol{\alpha}}_B \widetilde{\mathbf{P}}_B + \widetilde{\boldsymbol{\beta}}_B \right) + \mathbf{A}_D \mathbf{P}_D, \tag{2.48}$$

$$\underline{\mathbf{P}}_B^{re} \leq \mathbf{P}_B \leq \overline{\mathbf{P}}_B^{re}, \tag{2.49}$$

$$\textit{Constraints } (2.41). \tag{2.50}$$

The notations mentioned above can be found in the constraints (2.38)–(2.42), except (1) the constant matrix $\widetilde{\mathbf{A}}_B$ in the constraint (2.45), (2) the vectors $\underline{\mathbf{P}}_B^{re}$ and $\overline{\mathbf{P}}_B^{re}$ in the constraint (2.49), and (3) the vectors $\widetilde{\boldsymbol{\alpha}}_B$ and $\widetilde{\boldsymbol{\beta}}_B$ in the constraints (2.47)–(2.48), and (4) the error bound $\boldsymbol{\varepsilon}_B$ in the constraints (2.47) and (2.48). They will be elaborated below.

The constraint (2.45) is a compact form of (2.43), leading to the easy calculation of $\widetilde{\mathbf{A}}_B$ based on (2.43). Each element in $\underline{\mathbf{P}}_B^{re}$ (resp. $\overline{\mathbf{P}}_B^{re}$) can be determined by solving an LP where the objective function is the minimum (resp. maximum) corresponding tie-line power flow exchange adhering to the constraints (2.38)–(2.42). For $\widetilde{\boldsymbol{\alpha}}_B$ and $\widetilde{\boldsymbol{\beta}}_B$, they are determined in the manner from in Appendix 2.6.1. As claimed in Appendix 2.6.1, the following property holds:

$$-\boldsymbol{\varepsilon}_B \leq \mathbf{A}_B \mathbf{P}_B - \left(\widetilde{\boldsymbol{\alpha}}_B \widetilde{\mathbf{P}}_B + \widetilde{\boldsymbol{\beta}}_B \right) \leq \boldsymbol{\varepsilon}_B, \tag{2.51}$$

where the non-negative error bound $\boldsymbol{\varepsilon}_B$ is an explicit function of $\widetilde{\boldsymbol{\alpha}}_B$ and $\widetilde{\boldsymbol{\beta}}_B$, as also calculated in Appendix 2.6.1.

For any point $\widetilde{\mathbf{P}}_B$ within $\widetilde{\mathbf{W}}$, there is a feasible $\mathbf{P}_G$ fulfilling the constraints (2.38)–(2.42). At the same time, the point $(\mathbf{P}_G, \mathbf{P}_B)$ within $\widetilde{\mathbf{Y}}$ also fulfill the constraints (2.38)–(2.42). In other words, $\widetilde{\mathbf{W}}$ from the constraints (2.45)–(2.50) can be identified as an inner approximation to the exact permissible region in the domain of $\widetilde{\mathbf{P}}_B$. ■

The proof of Proposition 2.1 is given in Appendix 2.6.2. Given Proposition 2.1, the permissible region can be represented in the domain of $\widetilde{\mathbf{P}}_{\mathrm{B}}$ instead of $\mathbf{P}_{\mathrm{B}}$, as shown in (2.44). In power industries, the number of border interactions between interconnected transmission sub-systems is much smaller than the number of tie-lines. Consequently, the dimension of $\widetilde{\mathbf{P}}_{\mathrm{B}}$ is much smaller than that of $\mathbf{P}_{\mathrm{B}}$, leading to a relaxed computational burden of identifying the permissible region in the domain of $\widetilde{\mathbf{P}}_{\mathrm{B}}$.

2.3.3 *Case Studies*

2.3.3.1 Numerical Experiments for the Acceleration Method in Sect. 2.3.1

Section 2.3.3.1 will examine the effectiveness of the acceleration method in Sect. 2.3.1 for a relaxed computational burden of resource aggregation in large-scale transmission sub-systems. The acceleration method will be verified in the IEEE 118-bus test system and a real-world 661-bus utility system, whose parameters have been uploaded to [15]. Particularly, the acceleration method in Sect. 2.3.1 will be tested in the following four scenarios:

S1: Two AC tie-lines are connected at bus 5 and bus 60 in the IEEE 118-bus test system, respectively.

S2: Two AC-tie lines are connected at bus 5 and bus 60, and a DC tie-line is connected at bus 20 in the IEEE 118-bus test system, respectively.

S3: Two AC tie-lines are connected at bus 5 and bus 7 in the real-world 661-bus utility system, respectively.

S4: Two AC tie-lines are connected at bus 5 and bus 7, and a DC tie-line is connected at bus 9 in the real-world 661-bus utility system, respectively.

Note that coupling through DC tie-lines only includes tie-line power flow exchanges, while coupling through AC tie-liens additionally includes border phase angles [18]. For the sake of simplicity, P_i is denoted as the tie-line power flow exchanges exported from/imported to border bus i, while θ_i is denoted as border phase angles at border bus i. Figures 2.9 and 2.10 are used to validate the computational accuracy of the permissible region of the acceleration method in Sect. 2.3.1. The benchmark is selected as the permissible region of the resource aggregation method in Sect. 2.2.2, which has been proven accurate in Sect. 2.2.3. As shown in Figs. 2.9 and 2.10, the permissible regions of the acceleration method in Sect. 2.3.1 are the same as those of the resource aggregation method in Sect. 2.2.2.

The computational time of the methods in Sect. 2.3.1 and Sect. 2.2.2 is compared in Table 2.3. The acceleration method in Sect. 2.3.1 takes the least computational time of all the four scenarios. Particularly, the computational time of the acceleration method in Sect. 2.3.1 can reach up to over 30 times faster than that of the resource aggregation method in Sect. 2.2.2, while the computational time of the resource aggregation method in Sect. 2.2.2 under S4 cannot terminate within one hour. The

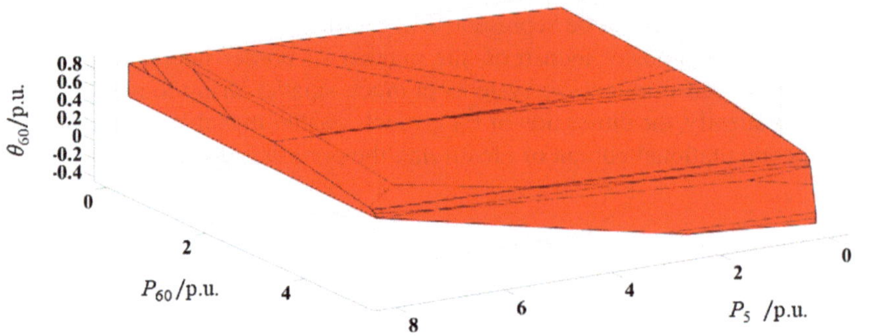

(a) Permissible region of the acceleration method in Sec. 2.3.1 under S1 in the IEEE 118-bus test system.

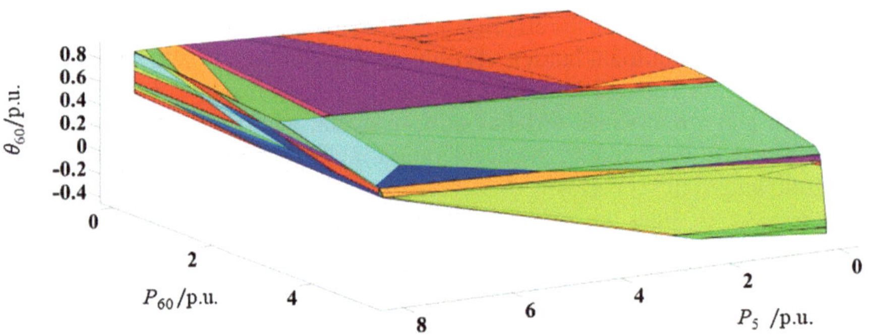

(b) Permissible region of the resource aggregation method in Sec. 2.2.2 under S1 in the IEEE 118-bus test system. Colors indicate different sub-regions found by traditional multi-parametric programming.

Fig. 2.9 Permissible regions in the IEEE 118-bus test system under different methods, reprinted from [13], with permission from IEEE

different computational time between the acceleration method in Sect. 2.3.1 and the resource aggregation method in Sect. 2.2.2 arises from their different computational burden, which is further compared in Table 2.4. Particularly, the computational burden is compared by counting the sub-region numbers and their computations for their final permissible regions. As shown in Table 2.4, the computational burden of the acceleration method in Sect. 2.3.1 is significantly less than that of the resource aggregation method in Sect. 2.2.2. This can be observed from the following two aspects:

- The number of found sub-regions. Compared with the resource aggregation method in Sect. 2.2.2, the total sub-region number of the acceleration method in Sect. 2.3.1 is reduced by 88.3, 84.24, 91.83 and 87.22% under S1–S4, respectively. The acceleration method in Sect. 2.3.1 improves traditional multi-parametric programming

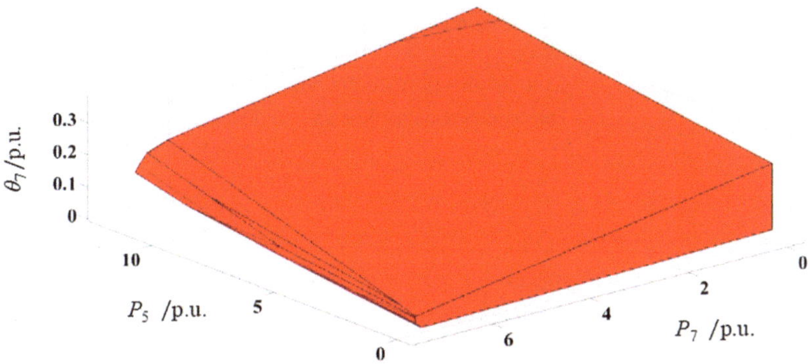

(a) Permissible region of the acceleration method in Sec. 2.3.1 under S1 in the 661-bus utility system.

(b) Permissible region of the resource aggregation method in Sec. 2.2.2 under S1 in the 661-bus utility system. Colors indicate different sub-regions found by traditional multi-parametric programming.

Fig. 2.10 Permissible regions in the 661-bus utility system under different methods, reprinted from [13], with permission from IEEE

by fixing the search direction along the boundary of the whole permissible region. Only sub-regions relevant to the boundary are required. In addition, the abundant search for sub-regions that share a boundary facet is avoided as well. However, the resource aggregation method in Sect. 2.2.2 has to search a lot of extra inner sub-regions and all the sub-regions on a shared boundary of the whole permissible region. Consequently, the total number of the found sub-regions of the acceleration method in Sect. 2.3.1 is significantly smaller than that of the resource aggregation method in Sect. 2.2.2.

- The number of sub-regions found based on optimization computations. The acceleration method in Sect. 2.3.1 adds a pre-processing stage before the pivot computation stage in traditional multi-parametric programming. The pre-processing stage provides the opportunity to find a sub-region by simple matrix manipulations instead of iterative optimization computations, while the pivot computation stage

Table 2.3 Computational time of different methods M1-M2 under different scenarios, reprinted from [13], with permission from IEEE

Scenario	Computational time (seconds)	
	The acceleration method in Sect. 2.3.1	The resource aggregation method in Sect. 2.2.2
S1	10.55	351.22
S2	47.69	1433.75
S3	44.63	1249.42
S4	200.46	5526.73

Table 2.4 Sub-region numbers of different methods under different scenarios, reprinted from [13], with permission from IEEE

Scenario	The acceleration method in Sect. 2.3.1			The resource aggregation method in Sect. 2.2.2
	N_{guess}	N_{pivot}	N_{Total1}	N_{Total2}
S1	10	52	62	530
S2	36	123	159	1009
S3	57	72	129	1578
S4	92	160	252	2889

Note N_{guess} is the number of sub-regions found by the pre-processing stage. N_{pivot} is the number of sub-regions found by the pivot computation stage. $N_{Total1} = N_{guess} + N_{pivot}$

repeats iterative optimization computations. However, all the found sub-regions in the resource aggregation method in Sect. 2.2.2 come from the pivot computation stage. Consequently, there is a smaller computational burden when sub-regions are found using the acceleration method in Sect. 2.3.1. For the acceleration method in Sect. 2.3.1, its percentages of sub-regions found by "fast guess" are 16.13, 22.64, 44.19 and 36.51% under S1–S4, respectively.

2.3.3.2 Numerical Experiments for the Acceleration Method in Sect. 2.3.2

Section 2.3.3.2 examines the effectiveness of the acceleration method in Sect. 2.3.2 for a relaxed computational burden under multiple border buses. It is verified in the IEEE 9-bus test system, whose parameters have been uploaded to [16]. There is no doubt that a smaller number of border buses indicates a smaller computational burden in the same system. Consequently, Sect. 2.3.3.2 will focus on the feasibility guarantee of the constraints (2.38)–(2.42) when employing the acceleration method in Sect. 2.3.2.

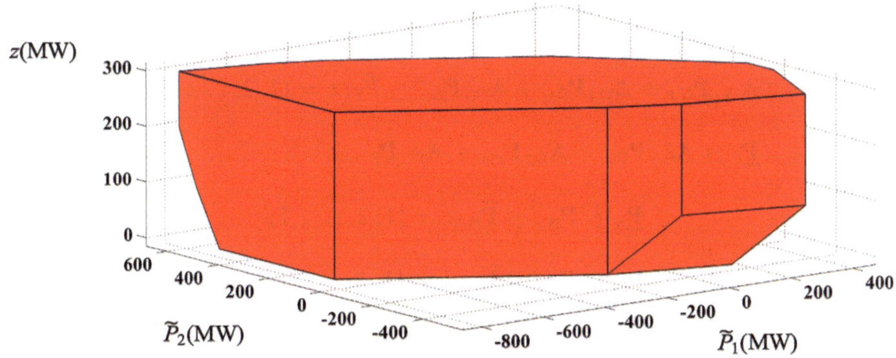

Fig. 2.11 Permissible region after bundled in Sect. 2.3.2, reprinted from [17], with permission from IEEE. $\widetilde{P}_1$ is the bundle of tie-lines at border buses 3 and 7. $\widetilde{P}_2$ is the bundle of tie-lines at border buses 1 and 9. Note that the z-axis implies the objective function (selected as the minimization of renewable curtailment in this testing) of one ED model

Four border buses 1, 3, 7, and 9 in the IEEE 9-bus test system are connected with four tie-lines, respectively. Four tie-lines are categorized into two groups. Tie-lines connected to border buses 1 and 9 are categorized as one group, while tie-lines connected to border buses 3 and 7 are categorized as another group. Furthermore, the permissible region of these two bundled tie-line power flow exchanges is calculated, as shown in Fig. 2.11. To examine the constraint violations of the permissible region in Fig. 2.11, 1×10^4 points are randomly selected from the permissible region. For each selected point, the feasibility of the transmission-level constraints (2.38)–(2.42) is checked. Numerical experiments report that all the selected points are feasible to the transmission-level constraints (2.38)–(2.42). In other words, the bundle of tie-line power flow exchanges in Sect. 2.3.2 can ensure the feasibility of the transmission-level constraints (2.38)–(2.42) with a relaxed computational burden under multiple border buses.

2.4 Temporal Acceleration-Oriented Resource Aggregation

Transmission sub-systems can operate in consecutive time periods, and thus intertemporal constraints are introduced in delineating the permissible region of tie-line power flow exchanges. This indicates the permissible region can contain multi-period tie-line power flow exchanges. This implies that the multi-period permissible region is indeed a high-dimension polytope. Consequently, identifying such a multi-period permissible region can face a heavy computational burden due to inherent boundaries and vertices for its delineation. This motivates accelerated computations in the context of multiple time periods to ensure the computational within limited operational time. For illustration purposes, the following intertemporal constraints will be considered as examples:

$$\mathbf{e}_{\mathrm{G},t}^{\top}\mathbf{P}_{\mathrm{G},t} + \mathbf{e}_{\mathrm{B},t}^{\top}\mathbf{P}_{\mathrm{B},t} = \mathbf{e}_{\mathrm{D},t}^{\top}\mathbf{P}_{\mathrm{D},t}, t = \{1, 2, \ldots, \mathrm{T}\}, \tag{2.52}$$

$$\mathbf{A}_{\mathrm{G},t}\mathbf{P}_{\mathrm{G},t} + \mathbf{A}_{\mathrm{B},t}\mathbf{P}_{\mathrm{B},t} + \mathbf{A}_{\mathrm{D},t}\mathbf{P}_{\mathrm{D},t} \leq \overline{\mathbf{P}}_{\mathrm{F}}, t = \{1, 2, \ldots, \mathrm{T}\}, \tag{2.53}$$

$$\underline{\mathbf{P}}_{\mathrm{F}} \leq \mathbf{A}_{\mathrm{G},t}\mathbf{P}_{\mathrm{G},t} + \mathbf{A}_{\mathrm{B},t}\mathbf{P}_{\mathrm{B},t} + \mathbf{A}_{\mathrm{D},t}\mathbf{P}_{\mathrm{D},t}, t = \{1, 2, \ldots, \mathrm{T}\}, \tag{2.54}$$

$$\underline{\mathbf{P}}_{\mathrm{B}} \leq \mathbf{P}_{\mathrm{B},t} \leq \overline{\mathbf{P}}_{\mathrm{B}}, t = \{1, 2, \ldots, \mathrm{T}\}, \tag{2.55}$$

$$\underline{\mathbf{P}}_{\mathrm{G}} \leq \mathbf{P}_{\mathrm{G},t} \leq \overline{\mathbf{P}}_{\mathrm{G}}, t = \{1, 2, \ldots, \mathrm{T}\}, \tag{2.56}$$

$$\underline{\mathbf{R}}_{\mathrm{G}} \leq \mathbf{P}_{\mathrm{G},t} - \mathbf{P}_{G,t-1} \leq \overline{\mathbf{R}}_{\mathrm{G}}, t = \{1, 2, \ldots, \mathrm{T}\}, \tag{2.57}$$

where all the notation explanations in the constraints (2.52)–(2.57) can be found in the constraints (2.38)–(2.42), except the ramp-down rates $\underline{\mathbf{R}}_{\mathrm{G}}$ and the ramp-up rates $\overline{\mathbf{R}}_{\mathrm{G}}$.

The topic of this section is to efficiently identify the following permissible region of multi-period tie-line power flow exchanges $\mathbf{P}_{\mathrm{B}}$ where $\mathbf{P}_{\mathrm{B}} = \left(\mathbf{P}_{\mathrm{B},1}, ..., \mathbf{P}_{\mathrm{B,T}}\right)$:

$$\mathbf{W} \triangleq \left\{\mathbf{P}_{\mathrm{B}} \mid \exists \left(\mathbf{P}_{\mathrm{G},1}, \ldots, \mathbf{P}_{\mathrm{G,T}}\right) \text{ fulfilling (2.52)–(2.57)}\right\}. \tag{2.58}$$

For different identification focuses, two acceleration methods will be introduced. The first method will focus on performance preservation (i.e., keeping the utilization potential of the exact permissible region under multiple time periods), as will be introduced in Sect. 2.4.1. The second method focuses on the feasibility guarantee, i.e., finding an inner approximation to the exact permissible region at the expense of performance loss, as will be introduced in Sect. 2.4.2.

2.4.1 *Performance Preservation-Oriented Acceleration*

2.4.1.1 Accelerated Computation

The basic idea is to approximate the original high-dimension polytope $\mathbf{W}$ exactly capturing the multi-period permissible region using an intersection of several lower-dimension polytopes that reflect time coupling among certain time periods. Consequently, the first issue in this method is to determine which periods are reflected in the ith lower-dimension polytope $\mathbf{W}_i$. To observe the original time coupling, all the possible combinations of time slots should be enumerated. However, this indicates that the total number of lower-dimension polytopes is $\mathrm{M} = C_{\mathrm{T}}^{2} + \ldots + C_{\mathrm{T}}^{\mathrm{T}-1} = 2^{\mathrm{T}} - 2 - \mathrm{T}$, where $C_{\mathrm{T}}^{\mathrm{j}}$ is the combination number when j time periods are selected from T time periods. The resultant value of M is large and thus too many lower-dimension polytopes should be identified. This additionally brings a considerable computational burden even if the original high-dimension is approximated using an intersection of several low-dimension polytopes. To avoid this issue, an alleviation strategy is further introduced. The strategy combines k ($2 \leq k \leq T$) arbitrary time periods. In this case, the number of required lower-dimension polytopes is $m = C_T^k$, which is much smaller than $M = 2^T - 2 - T$. Then, identifying the high-dimension polytope $\mathbf{W}$ only requires the identifications of lower-dimension polytopes $\mathbf{W}_i$ ($i = 1, 2, \ldots, m$), where m is the total number of the lower-dimension polytopes). For each lower-dimension polytope, it can be efficiently identified by multi-parametric programming. Appendix 2.6.3 theoretically analyzes the computational complexity of this acceleration method and the resource aggregation method in Sect. 2.2.2 which directly employs traditional multi-parametric programming. Based on the analysis from Appendix 2.6.3, the acceleration effectiveness of this acceleration method can be observed theoretically.

2.4.1.2 Performance Evaluation

For convenience of discussions, let's define the final result from Sect. 2.4.1.1 as $\mathbf{GP}_B \leq \mathbf{F}$ where the matrix $\mathbf{G}$ and the vector $\mathbf{F}$ are constant after implementing the acceleration method in Sect. 2.4.1.1. One coming problem is to evaluate how much performance will be preserved in $\mathbf{GP}_B \leq \mathbf{F}$. To achieve this goal, a performance evaluation method will be introduced below.

The multi-period permissible region of tie-line power flow exchanges is essentially a projection of the constraints (2.52)–(2.57) from the domain ($\mathbf{P}_G, \mathbf{P}_B$) to a tie-line domain $\mathbf{P}_B$. Consequently, the constraints (2.52)–(2.57) can be directly used as the benchmark to evaluate the performance preserved in $\mathbf{GP}_B \leq \mathbf{F}$. Note that both of the constraints (2.52)–(2.57) and $\mathbf{GP}_B \leq \mathbf{F}$ describe convex polytopes because of their linear formats. For an LP problem, its optimal solutions are always a vertex of the convex polytope delineated by its linear constraints. Also, polytopes delineated by two non-equivalent constraints usually contain different vertices. This provides the opportunity to evaluate the performance of $\mathbf{GP}_B \leq \mathbf{F}$ by employing different hyperplanes to scan polytopes delineated by the constraints (2.52)–(2.57) and $\mathbf{GP}_B \leq \mathbf{F}$ to find vertices, as shown in Fig. 2.12.

Based on the linear optimization theory, the following two LP problems are constructed under a given hyperplane $\mathbf{D}^\top \mathbf{P}_B$ which scans the polytope associated with either the linear constraints (2.52)–(2.57) or $\mathbf{GP}_B \leq \mathbf{F}$:

$$g_O^* = \min_{\mathbf{P}_G, \mathbf{P}_B} \mathbf{D}^\top \mathbf{P}_B \quad \text{s.t. Constraints (2.52)–(2.57)}, \tag{2.59}$$

$$g_A^* = \min_{\mathbf{P}_B} \mathbf{D}^\top \mathbf{P}_B \quad \text{s.t. } \mathbf{GP}_B \leq \mathbf{F}, \tag{2.60}$$

where $\mathbf{D}$ is the coefficient vector which describes a certain relationship among tie-line power flow exchanges in the LP problems (2.59) and (2.60). The objective function in the LP problems (2.59) and (2.60) indicates the maximum tie-line power flow exchanges adhering to the certain relationship restricted by the coefficient vector $\mathbf{D}$.

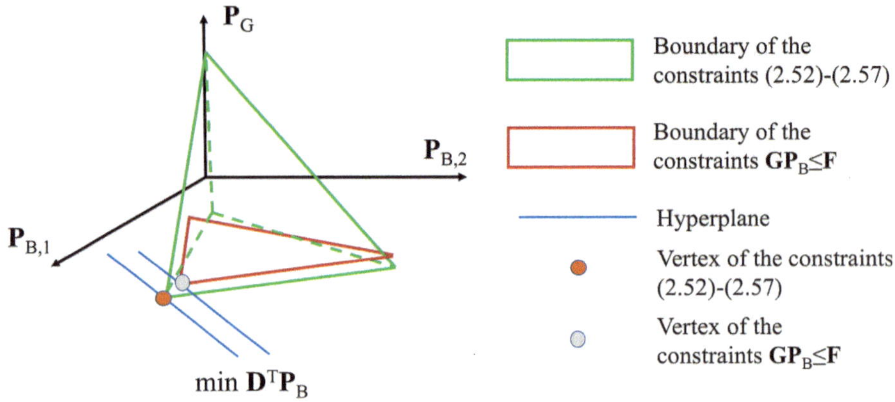

Fig. 2.12 Graphic illustration of the performance evaluation method, reprinted from [18], with permission from IEEE

For the same objective functions $\mathbf{D}^\top\mathbf{P}_B$ (i.e. the same hyperplane in Fig. 2.12), if the constraints (2.52)–(2.57) can provide the vertices which are similar or even the same as those from $\mathbf{GP}_B \leq \mathbf{F}$, the performance of the permission region from the acceleration method in Sect. 2.4.1.1 is good. A sufficiently large number of different hyperplanes $\mathbf{D}^\top\mathbf{P}_B$ provide the opportunity to compare all the vertices that describe the convex polytopes associated with either the constraints (2.52)–(2.57) or $\mathbf{GP}_B \leq \mathbf{H}$. Also, elements in $\mathbf{D}$ should be randomly generated so as to randomly compare vertices in the convex polytopes. To obtain a large number of hyperplanes which are randomly generated, a Monte Carlo framework is appropriate for the above-mentioned performance evaluation method.

After implementing a Monte Carlo framework to repeatedly solve the LP problems (2.59) and (2.60), the following five evaluation indexes can be calculated:

(1) Maximum error e_f^M of the objective function. For N_S generated hyperplanes, e_f^M is the maximum error between all the optimal objective functions $g_{O,j}^*$ $(j = 1, 2, \ldots, N_S)$ in the LP problem (2.59) and those optimal objective functions $g_{A,j}^*$ $(j = 1, 2, \ldots, N_S)$ in the LP problem (2.60), i.e.,

$$e_f^M = \max_{j=1,2,\ldots,N_S} \left|g_{A,j}^* - g_{O,j}^*\right|. \tag{2.61}$$

The index e_f^M evaluates the performance by measuring the potential maximum error of the maximum tie-line power flow exchanges given random relationships (reflected by the random $\mathbf{D}$) among different tie-line power flow exchanges.

(2) Average maximum error e_f^A of the objective function. The index e_f^A evaluates the performance by measuring the average error of the summation of all the tie-line power flow exchanges (given random relationships reflected by the random $\mathbf{D}$) across all the multiple time periods, that is

$$e_{\mathrm{f}}^{\mathrm{A}} = \frac{\sum_{j=1}^{\mathrm{N_S}} \left| g_{\mathrm{A},j}^{*} - g_{\mathrm{O},j}^{*} \right|}{\mathrm{N_S}}, \tag{2.62}$$

(3) Maximum error $e_{\mathrm{w}}^{\mathrm{M}}$ of optimal solutions. The index $e_{\mathrm{w}}^{\mathrm{M}}$ measures the maximum error between all the tie-line power flow exchanges at the optimum of the LP problem (2.59) and those of the LP problem (2.60), i.e.,

$$e_{\mathrm{w}}^{\mathrm{M}} = \max_{\substack{t=1,2,\ldots,\mathrm{T} \\ i=1,2,\ldots,\mathrm{N_B} \\ j=1,2,\ldots,\mathrm{N_S}}} \left| P_{\mathrm{BA},t,ij}^{*} - P_{\mathrm{BO},t,ij}^{*} \right|, \tag{2.63}$$

where $P_{\mathrm{BA},t,ij}^{*}$ is the optimal solution of the LP problem (2.60) relevant to the jth tie-line power flow exchange at time period t; $P_{\mathrm{BO},t,ij}^{*}$ is the optimal solution of the LP problem (2.59) relevant to the jth tie-line at time period t.

The index $e_{\mathrm{w}}^{\mathrm{M}}$ evaluates the performance by measuring the potential maximum error tie-line power flow exchange at each border bus at a single period.

(4) Average maximum error $e_{\mathrm{w}}^{\mathrm{A}}$ of optimal solutions. The index $e_{\mathrm{w}}^{\mathrm{A}}$ evaluates the performance by measuring the average error of tie-line power flow exchanges at a single border node at a single period, that is

$$e_{\mathrm{w}}^{\mathrm{A}} = \frac{\sum_{j=1}^{\mathrm{N_S}} \max_{\substack{t=1,2,\ldots,\mathrm{T} \\ i=1,2,\ldots,\mathrm{N_B}}} \left| P_{\mathrm{BA},t,ij}^{*} - P_{\mathrm{BO},t,ij}^{*} \right|}{\mathrm{N_S}}. \tag{2.64}$$

(5) Coincident rate γ. The index γ evaluates the performance by measuring the shape similarity, i.e., the following coincident rate of the same optimal results in the LP problems (2.59) and (2.60):

$$\gamma = \frac{\mathrm{N_c}}{\mathrm{N_s}} \times 100\% \tag{2.65}$$

where $\mathrm{N_c}$ is the number of hyperplanes with which the optimal results of the LP problem (2.59) are the same as those of the LP problem (2.60).

2.4.2 Feasibility Guarantee-Oriented Acceleration

Although the acceleration method in Sect. 2.4.1 has the potential for performance preservation for the exact multi-period permissible region of tie-line power flow exchanges in (2.58), its certain tie-line power flow exchanges may not be feasible to the multi-period constraints (2.52)–(2.57). To avoid such constraint violations which may threaten the security of low-carbon power systems, Sect. 2.4.2 will introduce a feasibility guarantee-oriented acceleration method for efficiently identifying an inner approximation to the exact multi-period permissible region of tie-line power flow exchanges. The key idea is to adjust the upper and lower bounds of generation levels in (2.56) so that the intertemporal constraints (2.57) are inherently redundant. In this context, the multi-period permissible region can be approximated across each time period while all the fulfillment of the transmission-level constraints (2.52)–(2.57) can be ensured.

Let's first adjust the bounds $\left(\underline{\mathbf{P}}_{\mathrm{G}}, \overline{\mathbf{P}}_{\mathrm{G}}\right)$ in the constraints (2.55) to the dispatch levels $\left(\mathbf{P}_{\mathrm{G},t}^{\min}, \mathbf{P}_{\mathrm{G},t}^{\max}\right)$ which comes from the following LP problem:

$$\max_{\left\{\mathbf{P}_{\mathrm{G},t}, \mathbf{P}_{\mathrm{B},t}, \mathbf{P}_{\mathrm{G},t}^{\max}, \mathbf{P}_{\mathrm{G},t}^{\min}\right\}, \forall t} \sum_{t=1}^{\mathrm{T}} \mathbf{e}_{\mathrm{G}}^{\top}\left(\mathbf{P}_{\mathrm{G},t}^{\max}-\mathbf{P}_{\mathrm{G},t}^{\min}\right), \tag{2.66}$$

$$\text{s.t. Constraints (2.52)–(2.55)}, t=\{1,2, \ldots, \mathrm{T}\}, \tag{2.67}$$

$$\mathbf{P}_{\mathrm{G},t}^{\max}-\mathbf{P}_{\mathrm{G},t-1}^{\min} \leq \overline{\mathbf{R}}_{\mathrm{G}}, t=\{1,2, \ldots, \mathrm{T}\}, \tag{2.68}$$

$$\mathbf{P}_{\mathrm{G},t-1}^{\max}-\mathbf{P}_{\mathrm{G},t}^{\min} \leq -\underline{\mathbf{R}}_{\mathrm{G}}, t=\{1,2, \ldots, \mathrm{T}\}, \tag{2.69}$$

$$\underline{\mathbf{P}}_{\mathrm{G}} \leq \mathbf{P}_{\mathrm{G},t}^{\min} \leq \mathbf{P}_{\mathrm{G},t}^{\max} \leq \overline{\mathbf{P}}_{\mathrm{G}}, t=\{1,2, \ldots, \mathrm{T}\}. \tag{2.70}$$

Given the dispatch levels $\left(\mathbf{P}_{\mathrm{G},t}^{\min}, \mathbf{P}_{\mathrm{G},t}^{\max}\right)$ from the LP problem (2.66)–(2.70), the following Proposition 2.2 demonstrates how to do an inner approximation to the exact multi-period permissible region of tie-line power flow exchanges across each time period.

Proposition 2.2 *Suppose that (1) there is an operating region $\mathbf{Y}_t$ represented by the constraints (2.52)–(2.55) and $\mathbf{P}_{\mathrm{G},t}^{\max} \leq \mathbf{P}_{\mathrm{G},t} \leq \mathbf{P}_{\mathrm{G},t}^{\min}$. Let's define the following set $\widetilde{\mathbf{W}}_t$ at each time period t:*

$$\widetilde{\mathbf{W}}_t \triangleq \left\{\mathbf{P}_{\mathrm{B},t} \mid \exists \mathbf{P}_{\mathrm{G},t} \in \mathbf{Y}_t\right\}. \tag{2.71}$$

Then, $\widetilde{\mathbf{W}}=\widetilde{\mathbf{W}}_1 \times \cdots \widetilde{\mathbf{W}}_t \times \cdots \times \widetilde{\mathbf{W}}_T$ is an inner approximation to the exact multi-period permissible region of tie-line power flow exchanges in (2.58), where the operator $\times$ indicates a Cartesian product. ∎

The proof of Proposition 2.2 is given in Appendix 2.6.4. From Proposition 2.2, it can be known that the Cartesian product of the permissible regions of tie-line power flow exchanges from each $\mathbf{Y}_t$ can serve as an inner approximation to the exact multi-period permissible region of tie-line power flow exchanges in (2.58). Moreover, the delineation of the permissible region of tie-line power flow exchanges from each $\mathbf{Y}_t$ can be processed by multi-parametric programming at a single time period, accelerating transmission-level resource aggregation under multiple time periods.

2.4.3 *Case Studies*

2.4.3.1 Numerical Experiments for the Acceleration Method in Sect. 2.4.1

Section 2.4.3.1 examines the effectiveness of the performance preservation-oriented acceleration method in Sect. 2.4.1 for a relaxed computational burden when facing multiple time periods. The acceleration method in Sect. 2.4.1 is verified in the following IEEE 118-bus test system and a real-world 661-bus utility system [19]:

S1: Two tie-lines are connected at border bus 5 and bus 9 in the IEEE 118-bus test system, respectively.

S2: Three tie-lines are connected at border bus 5 and bus 9 and bus 60 in the real-world 661-bus utility system, respectively.

Particularly, numerical experiments will exhibit the computational burden and performance evaluation results before and after implementing the acceleration method in Sect. 2.4.3.1. The computational time and performance evaluation results are listed in Tables 2.5 and 2.6, respectively.

- Analysis for computational time. Identifying the exact multi-period permissible region of tie-line power flow exchanges is time-consuming due to the explosion of its vertex number. Consequently, the exact permissible region cannot be identified before implementing the acceleration method in Sect. 2.4.1 within one hour. If the acceleration method in Sect. 2.4.1 is implemented, the computational time can significantly decrease to 96.96 s in the scenario S1 and 14.07 s in the scenario S2, as shown in Table 2.5. This is because only a few polytopes associated with combinations of different time periods are required. As theoretically analyzed in Appendix 2.6.3, the complexity of identifying a multi-period permissible region can significantly decrease if the acceleration method in Sect. 2.4.1 is implemented, pointing out the reasons why the computational time can decrease with the acceleration method in Sect. 2.4.1. It should be noted that the identification of each lower-dimension polytope is independent. Consequently, the computational efficiency of the acceleration method in Sect. 2.4.1 can be further improved by a parallel-computing framework, i.e., computing of each lower-dimension polytope in parallel.
- Analysis for performance evaluation. As shown in Table 2.6, the indexes $e_{\mathrm{f}}^{\mathrm{M}}$, $e_{\mathrm{f}}^{\mathrm{A}}$, $e_{\mathrm{w}}^{\mathrm{M}}$, and $e_{\mathrm{f}}^{\mathrm{A}}$ exhibit rather small values if the acceleration method in Sect. 2.4.1 is

Table 2.5 Computational time before and after implementing the acceleration method in Sect. 2.4.1 under different scenarios, reprinted from [18], with permission from IEEE

Scenario	Before implementing the acceleration method in Sect. 2.4.1 (s)	After implementing the acceleration method in Sect. 2.4.1 (s)
S1	$>3.6 \times 10^3$	96.96
S2	$>3.6 \times 10^3$	14.07

Table 2.6 Performance evaluation after implementing the acceleration method in Sect. 2.4.1 under different scenarios, reprinted from [18], with permission from IEEE

Scenario	$e_{\rm f}^{\rm M}$(p.u.)	$e_{\rm f}^{\rm A}$(p.u.)	$e_{\rm w}^{\rm M}$(p.u.)	$e_{\rm f}^{\rm A}$(p.u.)	γ (%)
S1	0.0012	3.3610×10^{-6}	0.5008	1.5945×10^{-4}	98.92
S2	0.1075	0.0088	0.3470	0.0027	97.44

implemented. As explained in Sect. 2.4.1, such smaller values imply better performance of the approximation to the exact multi-period permissible region. Another important index is the coincident rate γ because this index describes a similarity degree between the approximated and the exact permissible regions of tie-line power flow exchanges. A high value of the index γ means that the approximation is more similar to the exact permissible region. As shown in Table 2.6, the indexes γ in the two scenarios always remain at high values when the acceleration method in Sect. 2.4.1 is employed. They can reach up to 95.76 and 97.44% in the scenario S1 and the scenario S2, respectively.

2.4.3.2 Numerical Experiments for the Acceleration Method in Sect. 2.4.2

Section 2.4.3.2 examines the effectiveness of the feasibility guarantee-oriented acceleration method in Sect. 2.4.2 for a relaxed computational burden when facing multiple time periods. It is verified in the IEEE 9-bus test system with two time periods, whose parameters have been uploaded to [16]. There is no doubt that the identification of a single-period permissible region is much faster than that of the multi-period permissible region, as somehow examined in Sect. 2.4.3.1 as well. Consequently, Sect. 2.4.3.2 majorly focuses on the feasibility guarantee of the transmission-level constraints (2.52)–(2.56) when employing the acceleration method in Sect. 2.4.2. As claimed in Sect. 2.4.2, the whole permissible region across the time periods 1 and 2 is a Cartesian product of the permissible regions at each time period from Proposition 2.2. To verify the legality of such time decomposition, 1×10^4 points are randomly selected from the permissible regions at two time periods. For each point, the feasibility of the transmission-level constraints (2.52)–(2.57) is checked. It is found that time coupling is not violated in the presence of decomposing time periods by adjusting dispatch levels from the LP problem (2.66)–(2.70). This indicates that the feasibility can be guaranteed even if time periods are decomposed in Proposition 2.2. This is majorly owned by the adjustments of dispatch levels from the LP problem (2.66)–(2.70) to ensure the inherent redundancy of the intertemporal constraints (2.57). Consequently, one multi-period permissible region can be approximated across each time period when ensuring the fulfillment of all the transmission-level constraints (2.52)–(2.57).

2.5 Summary

In this chapter, the resource aggregation for low-carbon power systems at the transmission level is introduced. The key core is the identification of permissible regions of tie-line power flow exchanges, by which different transmission sub-systems interact. Particularly, each combination of tie-line power flow exchanges in one transmission sub-system is expected to be feasible to its transmission-level constraints.

In order to identify such a permissible region of tie-line power flow exchanges, the multi-parametric programming theory is introduced. The multi-parametric programming theory points out that each combination of active and inactive constraints of transmission-level constraints represents a sub-region of the whole permissible region when tie-line power flow exchanges are regarded as varying parameters. By implementing multi-parametric programming, the exact permissible region of tie-line power flow exchanges can be identified by computing the union of all the found sub-regions. However, the permissible region of tie-line power flow exchanges would be rather complex due to certain issues, where a lot of sub-regions may be required to delineate the complex boundary of the permissible region. For example, the increasing transmission-level constraints in large-scale transmission sub-systems lead to the explosive growth of constraint combinations. Another example is the high dimension of the permissible region arising from multiple border buses and multiple time periods. The computational burden arising from these either spatial or temporal issues is alleviated in this chapter by developing various specially designed acceleration methods.

For the spatial issues, the computational burden in large-scale transmission sub-systems is first alleviated by modifying multi-parametric programming. This is achieved by introducing various strategies by which the permissible region of tie-line power flow exchanges can be found by only exploring its sub-regions relevant to the boundary when iterative optimization computations in multi-parametric programming can be minimized as much as possible. Additionally, the computational burden under multiple border buses is alleviated by bundling tie-line power flow exchanges. By building tie-line power flow exchanges based on their geographic locations (i.e., border interface), the permissible region of tie-line power flow exchanges can be represented in the domain of bundled tie-line power flow exchanges, which reduces the computational burden by the dimension decrease of tie-line power flow exchanges.

For the temporal issues, the computational burden under multiple time periods is alleviated. Two acceleration methods are specially introduced for different purposes. The first acceleration method is introduced for performance preservation (i.e., keeping the utilization potential of the exact permissible region under multiple time periods), which approximates the multi-period permissible region of tie-line power flow exchanges using a few lower-dimension polytopes reflecting time coupling across different time periods. The second acceleration method is introduced for the feasibility guarantee (i.e., finding an inner approximation to the exact multi-period permissible region at the expense of performance loss), which decomposes time periods for the inherent redundancy of intertemporal constraints so that a multi-period permissible region can be approximated across each time period.

2.6 Appendixes

2.6.1 Computations for (2.51)

The lth element in $\mathbf{A}_\mathrm{B}\mathbf{P}_\mathrm{B}$ in the constraints (2.39) and (2.40) can be separated based on the groups of tie-line power flow exchanges, as described by

$$\mathbf{A}_\mathrm{B}^{l}\mathbf{P}_\mathrm{B} = \sum_{j=1}^{\mathrm{J}}\left(\sum_{i=1}^{n_j}\alpha_\mathrm{B}^{l,j_i}P_\mathrm{B}^{j_i}\right), \tag{2.72}$$

where $\mathbf{A}_\mathrm{B}^{l}$ is the lth-row sub-matrix of $\mathbf{A}_\mathrm{B}$; $\alpha_\mathrm{B}^{l,j_i}$ is the coefficient in $\mathbf{A}_\mathrm{B}^{l}$ associated with $P_\mathrm{B}^{j_i}$.

The term $\sum_{i=1}^{n_j}\alpha_\mathrm{B}^{l,j_i}P_\mathrm{B}^{j_i}$ is approximated by $\widetilde{\alpha}_\mathrm{B}^{l,j}P_\mathrm{B}^{j}+\widetilde{\beta}_\mathrm{B}^{l,j}$, where P_B^{j} is the aggregation of $\sum_{i=1}^{n_j}P_\mathrm{B}^{j_i}$ from (2.43). In other words, $\mathbf{A}_\mathrm{B}\mathbf{P}_\mathrm{B}$ in the constraints (2.39) and (2.40) can be approximated by $\widetilde{\boldsymbol{\alpha}}_\mathrm{B}\widetilde{\boldsymbol{P}}_\mathrm{B}+\widetilde{\boldsymbol{\beta}}_\mathrm{B}$ after computing $\widetilde{\alpha}_\mathrm{B}^{l,j}$ and $\widetilde{\beta}_\mathrm{B}^{l,j}$. The resultant approximation errors can be bounded by the bound $\boldsymbol{\varepsilon}_\mathrm{B}$. Particularly, the lth element in $\boldsymbol{\varepsilon}_\mathrm{B}$ is denoted as $\varepsilon_\mathrm{B}^{l}$ and it can computed as

$$\varepsilon_\mathrm{B}^{l} = \sum_{j=1}^{\mathrm{J}}\varepsilon_\mathrm{B}^{l,j}, \tag{2.73}$$

where $\varepsilon_\mathrm{B}^{l,j}$ is computed below.

$$\varepsilon_\mathrm{B}^{l,j} = \min_{\widetilde{\alpha}_\mathrm{B}^{l,j}\widetilde{\beta}_\mathrm{B}^{l}}\max_{P_\mathrm{B}^{j_i}}\left|\sum_{i=1}^{n_j}\left(\alpha_\mathrm{B}^{l,j_i}P_\mathrm{B}^{j_i}\right)-\left(\widetilde{\alpha}_\mathrm{B}^{l,j}\sum_{i=1}^{n_j}\left(P_\mathrm{B}^{j_i}\right)+\widetilde{\beta}_\mathrm{B}^{l,j}\right)\right|, \tag{2.74}$$

$$\text{s.t.}\quad \text{Constraints (2.49).} \tag{2.75}$$

Particularly, the solutions of the optimization problem (2.74) and (2.75) can be explicitly constructed by the following compact optimization problem:

$$\varepsilon_0 = \min_{a_0,b_0,x_s}\left|\sum_{s=1}^{\mathrm{S}}a_sx_s-\left(a_0\sum_{s=1}^{\mathrm{S}}x_s+b_0\right)\right|, \tag{2.76}$$

$$\text{s.t.}\quad 0\le x_s\le x_s^{\max},\ s=\{1,2,\ldots,\mathrm{S}\}. \tag{2.77}$$

where a_0, b_0 and x_s represent $\widetilde{\alpha}_{\mathrm{B}}^{l,j}$, $\widetilde{\beta}_{\mathrm{B}}^{l}$ and $P_{\mathrm{B}}^{j_i}$, respectively. Note that the lower bound x_s is zero. This can be achieved by a simple linear transformation.

If supposing that the coefficients in (2.76) are ranked by $a_1 \leq a_2 \leq \ldots \leq a_n$, the globally optimal solution of (2.76) and 2.77) can be explicitly formulated as follows (see Theorem 2 in [20]):

$$a_0^* = a_c, \tag{2.78}$$

$$b_0^* = \frac{1}{2}\sum_{s=1}^{S}\left(a_s - a_0^*\right)x_s^{\max}, \tag{2.79}$$

$$\varepsilon_0^* = \frac{1}{2}\left[\sum_{s=c}^{n} a_s x_s^{\max} - \sum_{s=1}^{c-1} a_s x_s^{\max} + a_0^*\left(\sum_{s=1}^{c-1} x_s^{\max} - \sum_{s=1}^{n} x_s^{\max}\right)\right], \tag{2.80}$$

where the index c is determined in the following way:

$$\lambda_c \leq 0,\ \lambda_{c+1} \geq 0, \tag{2.81}$$

where λ_c in (2.81) is denoted as

$$\lambda_c = \sum_{s=1}^{c-1} x_s^{\max} - \sum_{s=c}^{n} x_s^{\max}. \tag{2.82}$$

Based on (2.73)–(2.74) and (2.78)–(2.80), $\widetilde{\alpha}_{\mathrm{B}}^{l,j}$ and $\widetilde{\beta}_{\mathrm{B}}^{l,j}$ can be easily obtained to construct the constant $\widetilde{\boldsymbol{\alpha}}_{\mathrm{B}}$, $\widetilde{\boldsymbol{\beta}}_{\mathrm{B}}$ and $\boldsymbol{\varepsilon}_{\mathrm{B}}$ in Proposition 2.1.

2.6.2 Proof of Proposition 2.1

This proof is majorly completed by checking the fulfillment of the constraints (2.38)–(2.42) when $(\mathbf{P}_{\mathrm{G}}, \mathbf{P}_{\mathrm{B}})$ within $\mathbf{Y}$ (delineated by (2.45)–(2.50)) is given.

(1) The constraints (2.41) and (2.42) can be directly fulfilled because they are also incorporated into $\mathbf{Y}$ via the constraints (2.50).
(2) The existence of the constraint (2.38) can be checked based on the constraints (2.45) and (2.46). The constraint (2.45) is a compact form of (2.43) that describes the bundled tie-line power flow exchanges. Particularly, each tie-line only belongs to a group, yielding

$$\widetilde{\mathbf{e}}_{\mathrm{B}}^{\top}\widetilde{\mathbf{P}}_{\mathrm{B}} = \mathbf{e}_{\mathrm{B}}^{\top}\mathbf{P}_{\mathrm{B}}, \tag{2.83}$$

where $\widetilde{\mathbf{e}}_{\mathrm{B}}$ is an all-one vector associated with $\widetilde{\mathbf{P}}_{\mathrm{B}}$.
Substituting (2.83) into the constraint (2.38) yields the constraint (2.46).

(3) The constraint (2.42) is fulfilled because the parameters in the constraints (2.49) are determined when the constraints (2.42) hold.
(4) The constraints (2.39)–(2.41) are fulfilled due to the constraints (2.47)–(2.48) and (2.51). The constraints (2.51) can be expanded as follows:

$$\mathbf{A}_{\mathrm{B}}\mathbf{P}_{\mathrm{B}} - \boldsymbol{\varepsilon}_{\mathrm{B}} \leq \widetilde{\boldsymbol{\alpha}}_{\mathrm{B}}\widetilde{\mathbf{P}}_{\mathrm{B}} + \widetilde{\boldsymbol{\beta}}_{\mathrm{B}}, \tag{2.84}$$

$$\widetilde{\boldsymbol{\alpha}}_{\mathrm{B}}\widetilde{\mathbf{P}}_{\mathrm{B}} + \widetilde{\boldsymbol{\beta}}_{\mathrm{B}} - \boldsymbol{\varepsilon}_{\mathrm{B}} \leq \mathbf{A}_{\mathrm{B}}\mathbf{P}_{\mathrm{B}}. \tag{2.85}$$

Combing the constraints (2.47) and (2.84) yields:

$$\begin{aligned} &\mathbf{A}_{\mathrm{G}}\mathbf{P}_{\mathrm{G}} + \mathbf{A}_{\mathrm{B}}\mathbf{P}_{\mathrm{B}} + \mathbf{A}_{\mathrm{D}}\mathbf{P}_{\mathrm{D}} - \boldsymbol{\varepsilon}_{\mathrm{B}} \\ &\leq \mathbf{A}_{\mathrm{G}}\mathbf{P}_{\mathrm{G}} + \left(\widetilde{\boldsymbol{\alpha}}_{\mathrm{B}}\widetilde{\mathbf{P}}_{\mathrm{B}} + \widetilde{\boldsymbol{\beta}}_{\mathrm{B}}\right) + \mathbf{A}_{\mathrm{D}}\mathbf{P}_{\mathrm{D}} \leq \overline{\mathbf{P}}_{\mathrm{F}} - \boldsymbol{\varepsilon}_{\mathrm{B}} \end{aligned}, \tag{2.86}$$

i.e., the constraints (2.39) hold. Similarity, the fulfillment of the constraints (2.40) can be checked based on the constraints (2.48) and (2.85).

2.6.3 *Computational Complexity Under Multi-periods*

This appendix analyzes the computational complexity of different methods to identify the multi-period permissible region of tie-line power flow exchanges in Sect. 2.4. As claimed in Sect. 2.2, the identification of one permissible region comes from finding all the combinations of active and inactive constraints. Consequently, this complexity analysis discusses the worst case to enumerate all the combinations of active and inactive constraints arising from the constraints (2.52)–(2.57). For see of discussions, let's denote (1) the number of constraints at each time period in the constraints (2.52)–(2.57) is U, (2) the number of generators and tie lines at each time period as N, and 3) the total number of time periods as T.

When directly applying traditional multi-parametric programming to identifying the multi-period permissible region from the constraints (2.52)–(2.57), it is easy to know that the combination number of active and inactive constraints is the following A_0:

$$\mathrm{A}_0 = C_{\mathrm{T}\times\mathrm{U}}^{\mathrm{T}\times\mathrm{N}}. \tag{2.87}$$

As for the acceleration method in Sect. 2.4.1, the multi-period permissible region is approximated by $C_{\mathrm{T}}^{\mathrm{k}}$ lower-dimension polytopes, where k is the number of time periods in each lower-dimension polytope. The combination number of active and inactive constraints in each polytope is $C_{\mathrm{k}\times\mathrm{U}}^{\mathrm{k}\times\mathrm{N}}$. Consequently, the total number of constraint combinations of the acceleration method in Sect. 2.4.1 can be calculated as the following A_1:

$$\mathrm{A}_1 = C_{\mathrm{T}}^{\mathrm{k}} \times C_{\mathrm{k}\times\mathrm{M}}^{\mathrm{k}\times\mathrm{N}}. \tag{2.88}$$

Let's further compare the complexity between A_0 and A_1 in the remaining content. Considering $C_{U}^{N} = C_{U\text{-}1}^{N} + C_{U\text{-}1}^{N\text{-}1}$ from [21], A_0 can be reformulated in the following recursive manner:

$$\begin{aligned} A_0 &= C_{T\times U}^{T\times N} \\ &= C_{T\times U\text{-}1}^{T\times N} + C_{T\times U\text{-}1}^{T\times N\text{-}1} \\ &= C_{T\times U\text{-}1}^{T\times N} + C_{T\times U\text{-}2}^{T\times N\text{-}1} + C_{T\times U\text{-}2}^{T\times N\text{-}2} \\ &= \cdots \\ &= \sum_{l=1}^{L} C_{T\times U-l}^{T\times N-l+1} + C_{T\times U\text{-}L}^{T\times N\text{-}L}, \end{aligned} \tag{2.89}$$

where L=(T − k) × N − 1.

Considering $C_{U}^{N} = C_{U\text{-}1}^{N\text{-}1} + C_{U\text{-}2}^{N\text{-}1} + \cdots + C_{N\text{-}1}^{N\text{-}1}$ from [22], the item $C_{T\times U\text{-}L}^{T\times N\text{-}L} = C_{T\times U\text{-}L}^{k\times N+1}$ in (2.89) can be reformulated as follows:

$$\begin{aligned} C_{T\times U\text{-}L}^{k\times N+1} &= C_{T\times U\text{-}L\text{-}1}^{k\times N} + C_{T\times U\text{-}L\text{-}2}^{k\times N} + \cdots + C_{k\times U}^{k\times N} + \cdots + C_{k\times N}^{k\times N} \\ &= \sum_{q=1}^{T\times U\text{-}L\text{-}1\text{-}k\times U} C_{T\times U\text{-}L-q}^{k\times N} + C_{k\times U}^{k\times N} + \sum_{q=1}^{k\times U\text{-}k\times N} C_{k\times U-q}^{k\times N} \end{aligned}. \tag{2.90}$$

Based on (2.89) and (2.90), A_0 can be equivalently formulated as follows:

$$\begin{aligned} A_0 = &\sum_{l=1}^{T\times N\text{-}k\times N\text{-}1} C_{T\times U-l}^{T\times N-l+1} + \sum_{q=1}^{T\times U\text{-}L\text{-}1\text{-}k\times U} C_{T\times U\text{-}L-q}^{k\times N} \\ &+ \sum_{q=1}^{k\times U\text{-}k\times N} C_{k\times U-q}^{k\times N} + C_{k\times U}^{k\times N} \end{aligned}. \tag{2.91}$$

From (2.89)–(2.91), the following two issues can be observed to support $A_1 \ll A_0$ when k is set to a small positive value:

(1) The values of $C_{T\times U-l}^{T\times N-l+1}$ and $C_{T\times M\text{-}L-q}^{k\times N}$ in (2.91) are generally larger than $C_{k\times U}^{k\times N}$. Also, the value of $C_{k\times U-q}^{k\times N}$ is positive.
(2) The values of L and (T × M-L-1-k × M) are much larger than C_{T}^{k}.

2.6.4 *Proof of Proposition 2.2*

This proof will show that any tie-line power flow exchanges from $\widetilde{\mathbf{W}}$ is also within the exact multi-period permissible region of tie-line power flow exchanges in (2.58). This can be resolved by proving that $\mathbf{P}_{B} = (\mathbf{P}_{B,1}, \ldots, \mathbf{P}_{B,T}) \in \widetilde{\mathbf{W}}$ and $\mathbf{P}_{G} = (\mathbf{P}_{G,1}, \ldots, \mathbf{P}_{G,T}) \in \mathbf{Y}$ where $\mathbf{Y} = \mathbf{Y}_1 \times \cdots \mathbf{Y}_t \times \cdots \times \mathbf{Y}_{n_T}$ fulfill the transmission-level constraints (2.52)–(2.57). Since all the transmission-level constraints are incorporated into the operating region $\mathbf{Y}$ except the intertemporal constraints (2.57), the remaining proof is to show that $(\mathbf{P}_{\mathrm{B}}, \mathbf{P}_{\mathrm{G}}) \in (\widetilde{\mathbf{W}}, \mathbf{Y})$ satisfy the intertemporal constraints (2.57).

When $\mathbf{P}_{\mathrm{G},t}^{\max} \leq \mathbf{P}_{\mathrm{G},t} \leq \mathbf{P}_{\mathrm{G},t}^{\min}$, it can be known that

$$\mathbf{P}_{\mathrm{G},t}^{\min} - \mathbf{P}_{\mathrm{G},t-1}^{\max} \leq \mathbf{P}_{\mathrm{G},t} - \mathbf{P}_{\mathrm{G},t-1} \leq \mathbf{P}_{\mathrm{G},t}^{\max} - \mathbf{P}_{\mathrm{G},t-1}^{\min}, t = \{1, 2, \ldots, \mathrm{T}\}. \tag{2.92}$$

Since the dispatch levels $\left(\mathbf{P}_{\mathrm{G},t}^{\min} \mathbf{P}_{\mathrm{G},t}^{\max}\right)$ are determined from the LP problem (2.66)–(2.70), (2.92) can be expressed as follows:

$$\underline{\mathbf{R}}_{\mathrm{G}} \leq \mathbf{P}_{\mathrm{G},t}^{\min} - \mathbf{P}_{\mathrm{G},t-1}^{\max} \leq \mathbf{P}_{\mathrm{G},t} - \mathbf{P}_{\mathrm{G},t-1} \leq \mathbf{P}_{\mathrm{G},t}^{\max} - \mathbf{P}_{\mathrm{G},t-1}^{\min} \leq \overline{\mathbf{R}}_{\mathrm{G}}, t = \{1, 2, \ldots, \mathrm{T}\}. \tag{2.93}$$

Based on (2.93), it is not hard to know that any $(\mathbf{P}_{\mathrm{B}}, \mathbf{P}_{\mathrm{G}}) \in (\widetilde{\mathbf{W}}, \mathbf{Y})$ satisfy the intertemporal constraints (2.57).

References

1. Feng, C., Wang, Y., Kong, P.Y., et al.: Provision of contingency frequency services for virtual power plants with aggregated models. IEEE Trans. Smart Grid **14**(4), 2798–2811 (2022)
2. Kang, C., Chen, X., Xu, Q., et al.: Balance of power: toward a more environmentally friendly, efficient, and effective integration of energy systems in China. IEEE Power Energy Mag. **11**(5), 56–64 (2013)
3. Lin, W., Yang, Z., Yu, J., et al.: Toward fast calculation of probabilistic optimal power flow. IEEE Trans. Power Syst. **34**(4), 3286–3288 (2019)
4. Lin, W., Yang, Z., Yu, J., et al.: Determination of transfer capacity region of tie lines in electricity markets: theory and analysis. Appl. Energy **239**, 1441–1458 (2019)
5. Zhang, M., Yang, Z., Lin, W., et al.: Enhancing economics of power systems through fast unit commitment with high time resolution. Appl. Energy **281**, 116051 (2021)
6. Carrión, M., Arroyo, J.M.: A computationally efficient mixed-integer linear formulation for the thermal unit commitment problem. IEEE Trans. Power Syst. **21**(3), 1371–1378 (2006)
7. Pistikopoulos, E., Georgiadis, M., Dua, V.: Multi-parametric Programming: Theory, Algorithms and Applications. WileyVCH, Weinheim (2007)
8. Boyd, S., Vandenberghe, L.: Convex Optimization. Cambridge University Press (2012)
9. Ji, Y., Thomas, R.J., Tong, L.: Probabilistic forecasting of real-time LMP and network congestion. IEEE Trans. Power Syst. **32**(2), 831–841 (2017)
10. System parameters. https://figshare.com/articles/Case_data/6739271
11. Yu, J., Liu, J., Li, W., et al.: Limit preserving equivalent method of interconnected power systems based on transfer capability consistency. IET Gener., Transm. Distrib. **10**(14), 3547–3554 (2016)
12. Yan, B., Luh, P.B., Zheng, T., et al.: A systematic formulation tightening approach for unit commitment problems. IEEE Trans. Power Syst. **35**(1), 782–794 (2020)
13. Lin, W., Yang, Z., Yu, J., et al.: Tie-line power transmission region in a hybrid grid: fast characterization and expansion strategy. IEEE Trans. Power Syst. **35**(3), 2222–2231 (2019)
14. Jones, C.N., Kerrigan, E.C., Maciejowski, J.M.: Lexicographic perturbation for multiparametric linear programming with applications to control. Automatica **43**(10), 1808–1816 (2007)
15. Case_data. https://figshare.com/articles/Case_data_for_fast_characterization_and_expmasion_strategy/7901258

16. CASE_DATA_HD_TIELINES. https://figshare.com/articles/dataset/CASE_DATA_HD_TIELINES/17708111
17. Lin, W., Jiang, H., Jian, H., et al.: High-dimension tie-line security regions for renewable accommodations. Energy **270**, 126887 (2023)
18. Lin, W., Yang, Z., Yu, J., et al.: Tie-line security region considering time coupling. IEEE Trans. Power Syst. **36**(2), 1274–1284 (2020)
19. Casedata_MultiPeriod. https://figshare.com/articles/Casedata_MultiPeriod/9790754
20. Li, X., Zhai, Q., Guan, X.: Robust transmission constrained unit commitment: a column merging method. IET Gener., Transm. Distrib. **14**(15), 2968–2975 (2020)
21. On a recursive formula for binomial coefficients. https://www.math.tamu.edu/berko/teaching/preREU2011/example_paper.pdf
22. Bijective proof problems. http://math.mit.edu/rstan/bij.pdf

Chapter 3
Resource Aggregation for Low-Carbon Power Systems at the Distribution Level

3.1 Introduction

Recent years have observed the fast and continuous development of *distributed energy resources* (DERs) (e.g., rooftop PV and electric vehicles) in one distribution sub-system. For example, 167 GW of rooftop PV were installed globally during 2019–2021, whose resultant peak outputs can be higher than the joint peak consumption of France and Britain [1]. Observed from one transmission sub-system, the incorporation of such DERs into its interconnected distribution sub-system at the *point of common coupling* (PCC) transfers the role of distribution sub-systems from pure power consumers to flexible prosumers [2]. This can be achieved by the following two perspectives. Firstly, local demands in one distribution sub-system can be fulfilled by the local DERs (e.g., rooftop PV) in addition to power transfers from its interconnected transmission sub-system. Secondly, DERs enable flexible energy demands, such as discharging of electric vehicles during peak periods and charging during off-peak periods. Consequently, DERs pave one promising way towards the coordinated energy management between one distribution sub-system and its interconnected transmission sub-system through flexible PCC power flow exchanges.

Unfortunately, limited to the small-scale capacities of individual DERs, heavy communications among massive DERs, and different stakeholders of DERs and the transmission-subsystem operator, DERs cannot be directly dispatched for such a coordinated energy management [3]. To avoid the issues above for the coordinated energy management by DER dispatch in one distribution sub-system, one alternative manner is to aggregate DERs with its distribution sub-system as one unified entity [4], i.e., the resource aggregation at the distribution level to identify the permissible region of PCC power flow exchanges. This chapter will introduce this resource aggregation at the distribution level based on its implementation procedure. Since massive DERs are connected to nodes in one distribution sub-system, they will be first aggregated as one unified entity in the distribution sub-system to ease its subsequent aggregation

W. Lin et al., *Coordinated Energy Management for Low-Carbon Power Systems*,
https://doi.org/10.1007/978-981-96-7533-3_3

together with the distribution sub-system for the identification of the permissible region of PCC power flow exchanges. The former resource aggregation is called node-level resource aggregation and the latter resource aggregation is called system-level resource aggregation. Their combination completes the resource aggregation at the distribution level.

The remaining part of this chapter is organized below. Section 3.2 will introduce the node-level resource aggregation method. By leveraging the convexity of individual DER constraints, a zonotope-based resource aggregation method will be introduced to efficiently aggregate DERs at the same node considering the trade-off between the aggregation accuracy and its impacts on the subsequent system-level aggregation. Section 3.3 will introduce the system-level resource aggregation method to identify the permissible region of PCC power flow exchanges. By leveraging Brouwer's fixed-point theorem, this resource aggregation method can identify an inner convex sub-region of the permissible region of PCC power flow exchanges when inevitably facing non-convex constraints (e.g., non-convex power flow equations) in the distribution sub-system. Moreover, an exploration strategy will be introduced to ensure a desirable coverage of the exact permissible region using multiple inner convex sub-regions. Finally, the summary of this chapter will be provided in Sect. 3.4.

3.2 Node-Level Resource Aggregation

3.2.1 Problem Formulation

Let's define (1) there are $\mathrm{N_T}$ time periods, and (2) denote p_t as the power consumption from one power consumer that owns several DERs. Particularly, p_t should respect the following individual DER constraints:

(1) The power consumption p_t should not exceed its upper and lower bounds, i.e.,

$$p_{t,\min} \leq p_t \leq p_{t,\max}, t = \{1, \ldots, \mathrm{N_T}\}, \tag{3.1}$$

(2) The operating state of DERs may change with the power consumption, such as the remaining capacity of a battery, room temperature in an HVAC system, etc. The rate of change of the state is assumed as a linear combination of the current state, exogenous variables, and the current power, i.e.,

$$dx_t/dt = ax_t + bu_t + cp_t, t = \{1, \ldots, \mathrm{N_T}\}, \tag{3.2}$$

where a, b, and c denote dynamic characteristic parameters. For example, in an *energy storage* (ES) system, $a \leq 0$ can represent the self-discharge rate of energy; b and c can represent the conversion efficiency of exogenous variables and power to the system state, respectively. After discretization, the state constraints (3.2) can be expressed as

$$x_{t,\min} \leq ax_t + bu_t + cp_t \leq x_{t,\max}, t = \{1, \ldots, \mathrm{N_T}\}. \tag{3.3}$$

The constraint (3.1) and (3.3) can describe the feasible set of various DERs, such as storage loads, controllable loads, HVAC systems, etc. For ease of discussions, the constraint (3.1) and (3.3) are compacted as

$$\mathbf{AP} \leq \mathbf{b}, \tag{3.4}$$

where $\mathbf{P}$ stacks all the p_t; the constant matrix $\mathbf{A}$ and the constant vector $\mathbf{b}$ can be obtained from the constraints (3.1) and (3.3). Figure 3.1 shows the geometric representations of the constraints (3.4) for an ES system.

For the power consumers who (1) own the DERs delineated by the constraints (3.4), and (2) are connected to the same node, they can be aggregated as one power demand $\mathbf{w}$ in the following feasible region:

$$\mathbf{W} \triangleq \{\mathbf{w} \mid \mathbf{w} = \sum_{j}^{\mathrm{N_J}} \mathbf{P}_j, \mathbf{A}_j\mathbf{P}_j \leq \mathbf{b}_j\}, \tag{3.5}$$

where P_j is the power consumption from the jth consumer; $\mathbf{A}_j$ and $\mathbf{b}_j$ are the parameters to delineate the jth power consumer in the form (3.4).

Actually, the feasible region (3.5) can be obtained by implementing the *Minkowski Sum* (M-Sum) [5]. Figure 3.2 illustrates an example of the M-Sum of two DERs over two time periods.

The M-Sum of convex polytopes is proven as a convex polyhedron [5]. Although some studies have attempted to implement the M-Sum of convex polytopes using the convex hull method [6] and the projection method [7], they still face computational challenges in high-dimensional cases in (3.5). To resolve this issue, a zonotope-based resource aggregation method will be next introduced.

3.2.2 *Zonotope-Based Aggregation*

3.2.2.1 Zonotope Definition

A zonotope is a special type of convex polytope, characterized by its center symmetry. It enables the resource aggregation for the feasible region (3.5) by (1) minimizing memory and computational effort and (2) explicit formulation of the M-Sum for various feasible regions. Consequently, in the zonotope-based resource aggregation method, the feasible region delineated by (3.4) for one power consumer is first approximated by a zonotope.

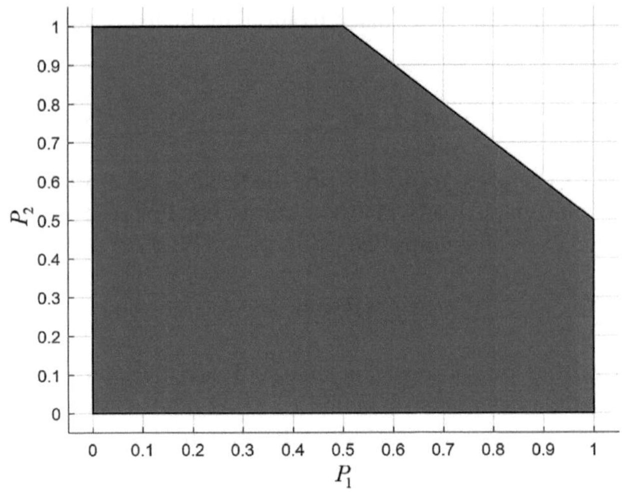

(a) Feasible region with two time periods.

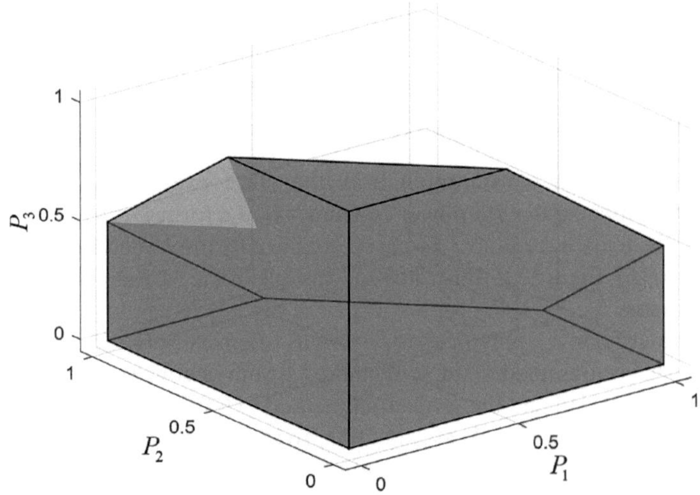

(b) Feasible region with three time periods.

Fig. 3.1 Feasible regions of an energy storage system in two-dimensional space, reprinted from [8], open access

The zonotope has the property of center symmetry, so it can be represented using a central point and multiple generators, i.e., delineated by the following feasible region:

$$\mathbf{Z} \triangleq \left\{ \mathbf{P} \mid \mathbf{P} = \mathbf{c} + \mathbf{G}\boldsymbol{\beta}, -\boldsymbol{\beta}_{\max} \le \boldsymbol{\beta} \le \boldsymbol{\beta}_{\max} \right\}, \tag{3.6}$$

where $\mathbf{c}$ denotes the center of the zonotope; $\mathbf{G}$ is the generator matrix in the form $\mathbf{G} = \left[\mathbf{g}_1, \ldots, \mathbf{g}_{\mathrm{N_G}}\right]$ where $\mathrm{N_G}$ is the number of generators.

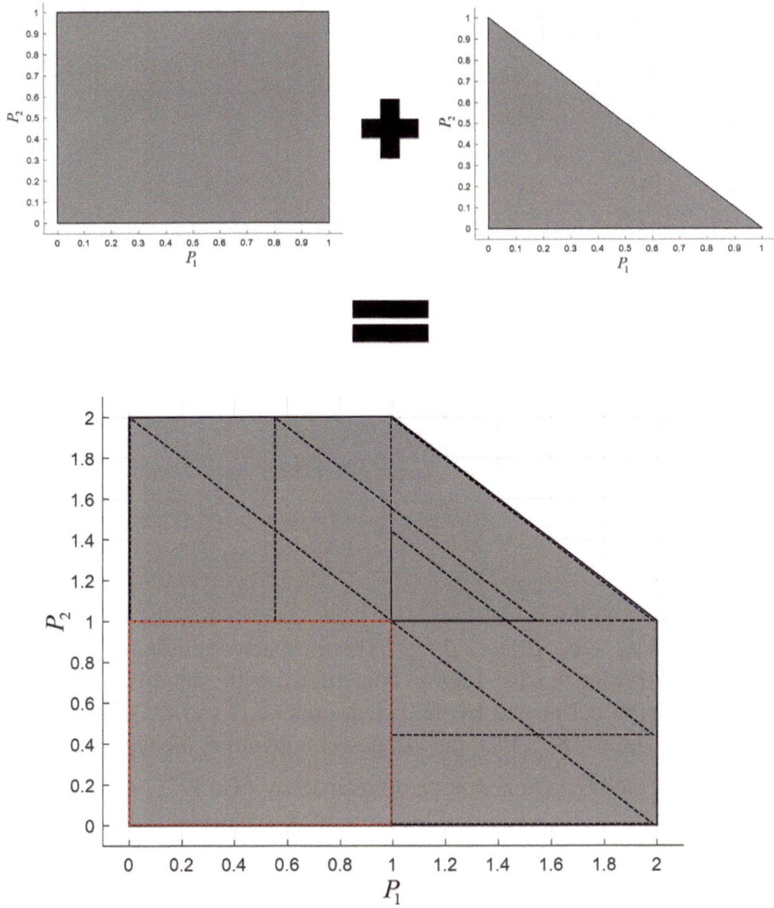

Fig. 3.2 Minkowski Sum of polytopes in a two-period decision cycle, reprinted from [8]

Particularly, an n-dimensional zonotope, the generators can be constructed as follows [9]:

$$\mathbf{g}^i = [0, \ldots, 0, 1, 0, \ldots, 0]^\top, \tag{3.7}$$

$$\mathbf{g}^{2n-1+i} = [0, \ldots, 0, -\frac{1}{\sqrt{2}}, \frac{1}{\sqrt{2}}, 0, \ldots, 0]^\top, \tag{3.8}$$

where the ith element in (3.7) equals one; the ith and $(i+1)$th elements in (3.8) equal $\frac{-1}{\sqrt{2}}$ and $\frac{1}{\sqrt{2}}$, respectively. Figure 3.3 shows a 2-D zonotope with 3 generators.

The zonotope (3.6) has an important property, namely that it is closed under Minkowski addition. For example, if one zonotope is delineated by $(\mathbf{G}, \mathbf{c}_1, \boldsymbol{\beta}_{\max}^{(1)})$ and another one zonotope is delineated by $(\mathbf{G}, \mathbf{c}_2, \boldsymbol{\beta}_{\max}^{(2)})$, their M-Sum is the zonotope

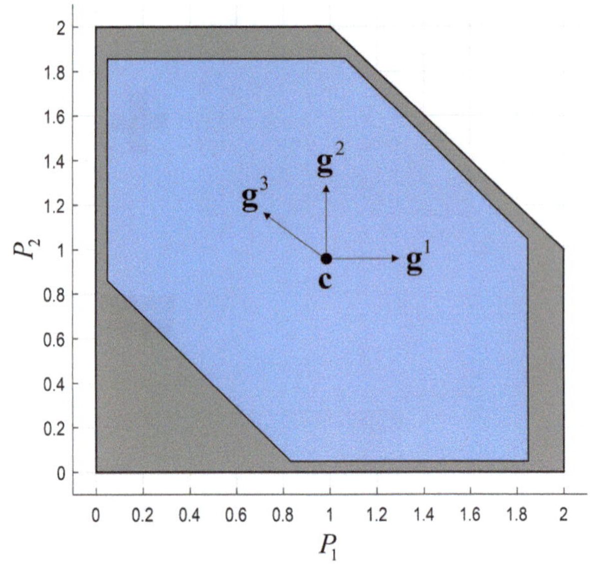

Fig. 3.3 2-D zonotope with three generators, reprinted from [8], open access

delineated by $\left(\mathbf{G}, \mathbf{c}_1 + \mathbf{c}_2, \boldsymbol{\beta}_{\max}^{(1)} + \boldsymbol{\beta}_{\max}^{(2)}\right)$. This property significantly enhances the efficiency of aggregating feasible regions constructed with zonotopes. In other words, if the feasible region delineated by the constraints (3.1) and (3.3) is approximated by one zonotope $\left(\mathbf{G}, \mathbf{c}_j, \boldsymbol{\beta}_{\max}^{(j)}\right)$ for the jth power consumer, the feasible region (3.5) can be also delineated by one zonotope delineated by $\left(\mathbf{G}, \sum_{j=1}^{\mathrm{N_J}} \mathbf{c}_j, \sum_{j=1}^{\mathrm{N_J}} \boldsymbol{\beta}_{\max}^{(j)}\right)$.

3.2.2.2 Zonotope-Based Approximation for (3.1) and (3.3)

Once the constraints (3.1) and (3.3) can be delineated by the zonotope (3.6), their M-Sum for the feasible region (3.5) is rather easy. Consequently, Sect. 3.2.2.2 tries to approximate the feasible region delineated by (3.1) and (3.3) as one zonotope while the approximation accuracy can be improved as much as possible. This approximation task will be solved by designing an optimization problem where the optimization objective is to maximize the similarity between the zonotope and the original feasible region, while ensuring an inner zonotope approximation.

First, if a zonotope is an inner approximation to the feasible region delineated by (3.1) and (3.3), $\mathbf{Ac} + |\mathbf{AG}|\boldsymbol{\beta}_{\max} \leq \mathbf{b}$ should hold [10]. Next, let's consider how to measure the similarity between the zonotope (3.6) and the feasible region delineated by (3.1) and (3.3). Intuitively, when the zonotope is an inner approximation to the feasible region delineated by (3.1) and (3.3), a larger volume of the zonotope indicates a higher approximation accuracy. However, the volume computation of a polytope requires its vertex representation, which has a heavy computational burden, especially in high-dimensional cases. Motivated by [11], the distance $\Delta_{Z,i}$ between each pair of parallel facets of the zonotope and the distance $\Delta_{P,i}$ between the two corresponding facets of the polytope are computed for similarity degree in an alternative manner, as geometrically demonstrated in Fig. 3.4.

By comparing the magnitude of $\Delta_{Z,i}$ and $\Delta_{P,i}$, the similarity degree between the zonotope and the feasible region delineated by (3.1) and (3.3) can be determined as

$$\Lambda_i = \frac{\Delta_{Z,i}}{\Delta_{P,i}} \tag{3.9}$$

where $\mathbf{N}_\mathrm{f}$ is the number of parallel facets in the zonotope. For the zonotope described by (3.7) and (3.8), $\mathrm{N}_\mathrm{f} = n^2 + n$. Let's denote $\Delta_Z = [\Delta_{Z,1}, \ldots, \Delta_{Z,\mathrm{N}_\mathrm{f}}]^\top$, with $\Delta_Z = 2|\mathbf{F}^\top \mathbf{G}|\boldsymbol{\beta}_{\max}$ where $\mathbf{F} = [f^1, \ldots, f^{\mathrm{N}_\mathrm{f}}]^\top$ is the matrix formed by the normal vectors of all the parallel facets of the zonotope. Let $\Delta_P = [\Delta_{P,1}, \ldots, \Delta_{P,\mathrm{N}_\mathrm{f}}]^\top$, and its calculation can be found in Algorithm 1.

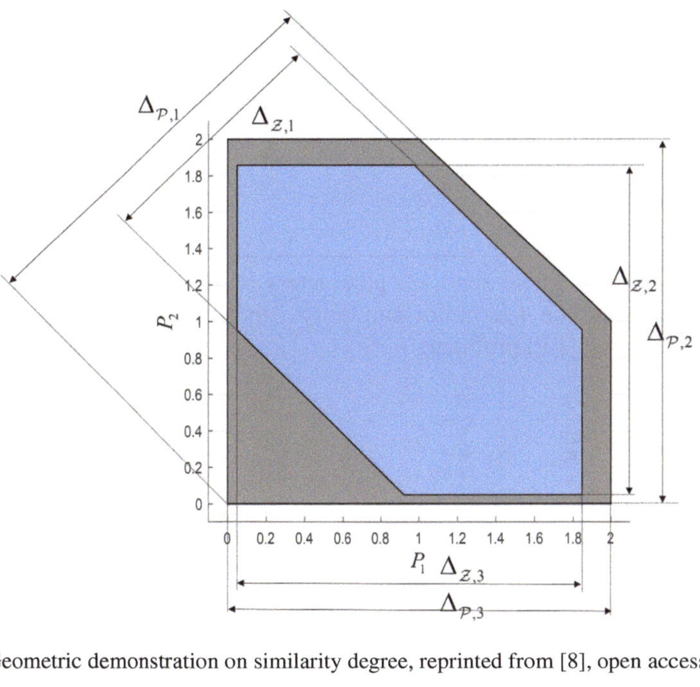

Fig. 3.4 Geometric demonstration on similarity degree, reprinted from [8], open access

Algorithm 1: Calculation of Δ_P of a Polytope

Input: Normal vector matrix $\mathbf{F}$ of the zonotope; $\mathbf{A}$, $\mathbf{b}$ of feasible region delineated by (3.1) and (3.3)
Output: Terminal distance matrix Δ_P
for *each normal vector* $\mathbf{f}^i$ *of each facet of the zonotope,* $i = 1, \ldots, N_f$ **do**
  Search the normal vector $\mathbf{a}^\mathbf{j}$ of the matrix $\mathbf{A}$ of the feasible region delineated by (3.1) and (3.3) that is linearly related to it;
  if *the number of normal vectors is 2* (i.e., $\mathbf{a}^i$, $\mathbf{a}^j$) **then**
    $\Delta_{P,1}$ is the Euclidean distance between the two parallel hyperplanes $\mathbf{a}^i\mathbf{P} = \mathbf{b}^i$ and $\mathbf{a}^j\mathbf{P} = \mathbf{b}^j$;
  else
    if *the number of normal vectors is 1* (i.e., $\mathbf{a}^i$) **then**
      Calculate the distance $\Delta_{a',v}$ from all the vertices $v \in V$ of the feasible region delineated by (3.1) and (3.3),
      $\Delta_{P,1} = \max\{\Delta_{a^i,v}\}$;
    else
      if *the number of normal vectors is 0* **then**
        Define the hyperplane $\mathbf{f}^i\mathbf{P} = \boldsymbol{\epsilon}$, and set $\boldsymbol{\epsilon}$ so that there is no intersection between the zonotope and the feasible region delineated by (3.1) and (3.3);
        Solve for $d_{\max}, d_{\min}$ using $d_{\max} = \max_{\mathbf{P}} \frac{\mathbf{f}^i\mathbf{P}-\boldsymbol{\epsilon}}{\|\mathbf{f}^i\|_2}$ s.t. $\mathbf{AP} \le \mathbf{b}$ and $d_{\min} = \min_{\mathbf{P}} \frac{\mathbf{f}^i\mathbf{P}-\boldsymbol{\epsilon}}{\|\mathbf{f}^i\|_2}$ s.t. $\mathbf{AP} \le \mathbf{b}$, $\Delta_{P,1} = d_{\max} - d_{\min}$;
      else
        –

After obtaining Δ_P, the zonotope parameters $(\mathbf{c}, \boldsymbol{\beta}_{\max})$ which ensures an inner approximation and the maximum similarity can be obtained by the following linear programming (LP) problem:

$$\max_{\mathbf{c},\boldsymbol{\beta}_{\max}} \quad \frac{1}{\mathrm{N_f}} \sum_{i=1}^{\mathrm{N_f}} \Lambda_i = \frac{2}{\mathrm{N_f}} (1./\Delta_P)^\top |\mathbf{F}^\top\mathbf{G}|\boldsymbol{\beta}_{\max} \tag{3.10}$$

$$\text{s.t.} \quad \mathbf{Ac} + |\mathbf{AG}|\boldsymbol{\beta}_{\max} \le \mathbf{b}, \tag{3.11}$$

$$\boldsymbol{\beta}_{\max} \ge \mathbf{0} \tag{3.12}$$

where ./ denotes element-wise division of matrix components.

3.2.2.3 Value-Oriented Approximation Improvements

Previously, the zonotope (3.6) is utilized to approximate the feasible region delineated by the constraints (3.1) and (3.3) for an efficient M-Sum. However, this improved efficiency is at the expense of the approximation error arising from the fact that the center symmetry of the zonotope cannot fully align with the feasible region delineated by the constraints (3.1) and (3.3). Consequently, one natural follow-up problem is how to minimize the impacts of the zonotope-based approximation on the node-level resource aggregation.

When clearing the support from the node-level resource aggregation in the current industries, it is generally divided into capacity costs and energy costs. The capacity costs are cleared based on the scale of the node-level resource aggregation results (i.e., the scale of the M-sum of various DERs), while the energy costs are calculated based on the actual energy usage, which changes with different time periods due to the clearing results. This means that the time value of the node-level resource aggregation results at different periods is different. Such a perspective can guide a more accurate zonotope-based approximation by a better fitting during high time-value periods. This will redistribute the inevitable approximation error to the time periods with lower time values; thus, it can ultimately enhance the value of the node-level resource aggregation results during the subsequent clearing process. This idea can be realized by assigning different weights to each component in Δ_Z and Δ_P, i.e.,

$$\Lambda_i' = \boldsymbol{\omega} \left(\left(\mathbf{f}^i \right)^2 \otimes \Lambda_i \right) \tag{3.13}$$

where $\boldsymbol{\omega}$ denotes the value weights of the node-level resource aggregation result at each time period; $(\mathbf{A})^2$ denotes the square of each element of the matrix $\mathbf{A}$; $\otimes$ denotes Hadamard product; Λ_i is the similarity degree from (3.9). This weight distribution manner (3.13) redistributes the distances $\Delta_{Z,i}$ and $\Delta_{P,i}$ between the zonotope and the feasible region delineated by (3.1) and (3.3) by (1) adjusting the decomposition direction of each coordinate, (2) and then assigning different weights to each component according to the value at each time period.

In summary, the value-oriented improvements on the zonotope-based approximation can be achieved by solving the following optimization problem:

$$\max_{\mathbf{c},\boldsymbol{\beta}_{\max}} \frac{1}{\mathrm{N_f}} \sum_{i=1}^{\mathrm{N_f}} \Lambda_i' = \frac{2}{\mathrm{N_f}} \left(\boldsymbol{\omega} \otimes \left(\mathbf{F}^2 \right) \right) \left((1./\Delta_P) \otimes |\mathbf{F}^\top \mathbf{G}| \boldsymbol{\beta}_{\max} \right) \tag{3.14}$$

$$\text{s.t. Constraints (3.11)–(3.12).} \tag{3.15}$$

The optimization problem (3.14) and (3.15) has the same computational complexity as the optimization problem (3.10)–(3.12), and they will have the same solutions when $\boldsymbol{\omega} = \mathbf{1}$.

Let's take Fig. 3.5 for an illustration of the optimization problem (3.14) and (3.15) under two time periods. In Fig. 3.5, the feasible region delineated by (3.1)–(3.3) and

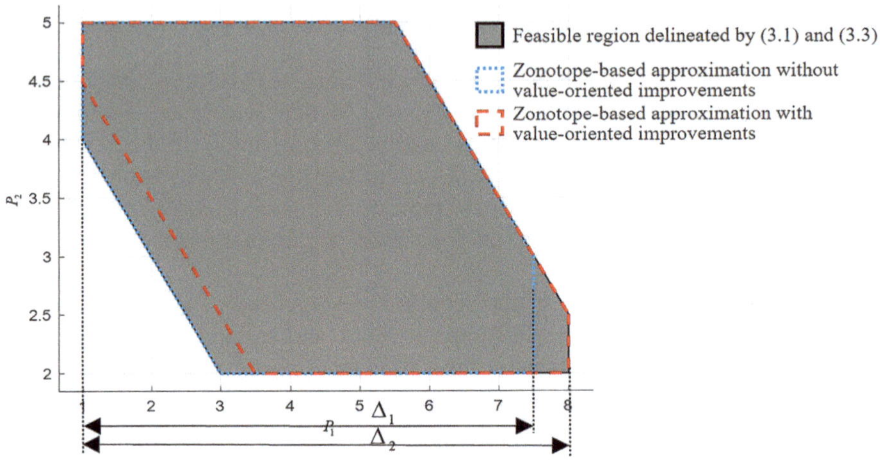

Fig. 3.5 Illustration of the optimization problem (3.14) and (3.15) under two time periods, reprinted from [8], open access

the zonotope approximation of the optimization problem (3.10)–(3.12) are represented using a grey box and blue dashed lines, respectively. Particularly, if the value at time period 1 is higher, let's set $\boldsymbol{\omega} = [2, 1]^\top$ by which the solutions of the optimization problem (3.14) and (3.15) will result in the zonotope approximation marked by the red solid lines in Fig. 3.5. As can be seen, Although the zonotope approximation without value considerations yields a larger shape, the zonotope approximation with value consideration is larger at time period t, i.e., $\Delta_2 > \Delta_1$. If capacity costs are taken into account, the result from the optimization problem (3.14) and (3.15) can yield a higher revenue.

3.2.3 *Case Studies*

3.2.3.1 Basic Information

The node-level resource aggregation method in Sect. 3.2 will be verified using three different DERs with different parameters and states, which are distributed ES systems [12]. Assuming these ES systems are equipped with ideal batteries (i.e., no charging or discharging losses with $a = b = 0, c = 1, x_k = \sum_{i=0}^{k} p_i$). The other parameters are listed in Table 3.1.

It is assumed that the aggregation of ES systems at the same node is a price taker and uses the price data from the Guangdong spot market in August 2020 as the value weights for value-oriented improvements.

Table 3.1 ES system parameters, reprinted from [8], open access

ES type	Maximum discharging power (kW)	Storage capacity (kWh)	Initial SoC
A	17.2	100	[0.2, 0.8]
B	50	210	[0.2, 0.8]
C	14	27	[0.2, 0.8]

Table 3.2 Similarity degree under different approximations and different ES systems, reprinted from [8], open access

Regulation period	ES type	Initial SoC	Similarity degree	
			Hyperrectangle-based approximation in [13]	Zonotope-based approximation in Sect. 3.2.2.2
4h	A	0.2	0.65	0.80
		0.5	0.73	0.91
		0.8	0.65	0.80
4h	B	0.2	0.55	0.79
		0.5	0.53	0.87
		0.8	0.55	0.79
4h	C	0.2	0.32	0.84
		0.5	0.25	0.86
		0.8	0.32	0.84
24h	A	0.2	0.12	0.52
		0.5	0.12	0.53
		0.8	0.12	0.52
24h	B	0.2	0.09	0.54
		0.5	0.09	0.55
		0.8	0.09	0.54
24h	C	0.2	0.05	0.69
		0.5	0.04	0.69
		0.8	0.05	0.69

3.2.3.2 Numerical Experiment Analysis

Let's first compare the effectiveness of the zonotope-based approximation in Sect. 3.2.2.2 to the feasible region delineated by (3.1) and (3.3). The regulation periods are set to 4h and 24h with different types and initial SoC. The approximation effectiveness is selected as the similarity degree from (3.10). To examine the superiority of the zonotope-based approximation, the hyperrectangle-based approximation in [13]. Numerical experiments are reported in Table 3.2.

Table 3.3 Computational time of the node-level resource aggregation, reprinted from [8], open access

Form to delineate one ES system	# of ES systems	Regulation period (h)	Time for an approximation (s)	Time for M-Sum (s)	Total time (s)
Equations (3.1) and (3.3)	10	4	–	410	410
Zonotope	10	4	43	0.03	43
Equations (3.1) and (3.3)	10	24	–	–	–
Zonotope	10	24	2425	0.04	2425
Zonotope	50	24	12355	0.04	12355

The following two findings can be observed from Table 3.2:

- For the zonotope-based approximation in Sect. 3.2.2.2, the average similarity degrees $\in [0, 1]$ for the 4h-regulation period and the 24h-regulation period are 0.83 and 0.58, respectively. This is because as the dimensionality increases with the regulation periods, the feasible region delineated by (3.1) and (3.3) becomes more complex, and thus approximation accuracy (i.e., the similarity degree) may decrease. For different types with the same regulation period, the parameters can significantly affect the similarity degree. The larger values of the maximum discharging power and storage capacity indicate the lower similarity degrees. For the same type with different initial SoC values, it is found that the initial SoC values significantly affect the similarity degree, and the highest similarity degree is observed when the initial SoC value is 0.5.
- For the hyperrectangle-based aggregation, its similarity degrees significantly drop compared with those of the zonotope-based aggregation. Particularly, the average similarity degrees of the 4h-regulation period and the 24h-regulation period drop to 0.5 and 0.09, respectively.

Although the zonotope-based approximation inevitably faces approximate error which increases with the increasing number of regulation periods, its approximation accuracy is significantly higher than that of the hyperrectangle-based approximation. Further, the computational time of the node-level resource aggregation based on the zonotope-based approximation in Sect. 3.2.2.2 is listed in Table 3.3.

When the regulation period is 4h and the number of ES systems is 10, the computational time for the M-Sum based on the feasible region delineated by (3.1) and (3.3) is 410 s, while the corresponding time based on the zonotope-based approximation is only 43 s, i.e., the calculation efficiency is improved by 9.5 times due to the employment of a zonotope-based approximation. When the regulation period is extended to 24h, the M-Sum based on the feasible region delineated by (3.1) and (3.3) cannot be computed, while the zonotope-based approximate facilitates the computation within 0.04 s for 10 and 50 ES systems. Although additional computational time is required to compute a zonotope-based approximation, such a computation can be implemented offline with a relaxed computational burden in online applications.

Based on aforementioned analysis, it is evident that the employment of a zonotope-based approximation can significantly improve the computational speed of the node-level resource aggregation with tremendous DERs, while the accuracy is somehow sacrificed. The remaining numerical experiments will examine the value-oriented improvements in Sect. 3.2.2.3 for an improved zonotope-based approximation. Particularly, the *maximum flexibility capacity ratio* (MFCR) of each ES system, i.e., the value of the projection of Δ_Z projected onto the axis-aligned normal vector, will be selected as an index to compare the resource aggregation results before and after implementing the value-oriented improvements in Sect. 3.2.2.3. When the value weights in the optimization problem (3.14)–(3.15) are selected as the prices from August 16th and August 30th in the data from the Guangdong province in China, the resultant numerical experiments are reported in Figs. 3.6 and 3.7. Figures 3.6 and 3.7 demonstrate that when the price fluctuations are in the middle of the day, the implementation of the value-oriented improvements in Sect. 3.2.2.3 can better provide a higher MFCR. As shown in Fig. 3.7, the MFCR change is inertial after implementing the value-oriented improvements in Sect. 3.2.2.3. This means that over long time periods with relatively low prices (T02-T10), the MFCR decreases gradually. When a price spike occurs (T18), the MFCR will not suddenly drop.

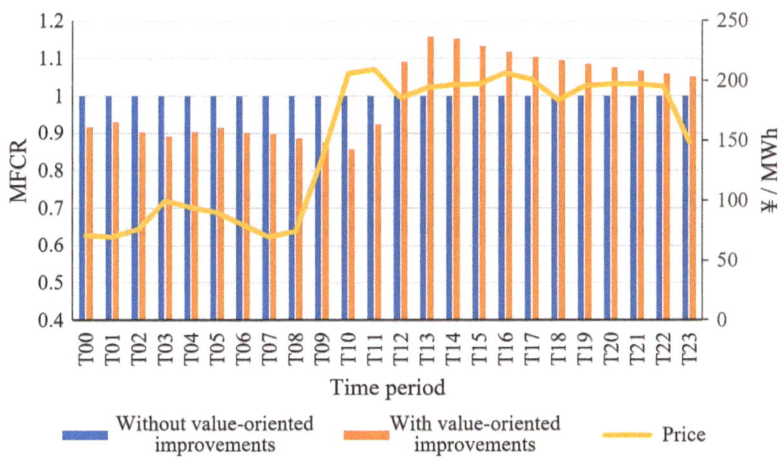

Fig. 3.6 MFCR comparison before and after the implementation of the value-oriented improvements in Sect. 3.2.2.3 (Aug 16), reprinted from [8], open access

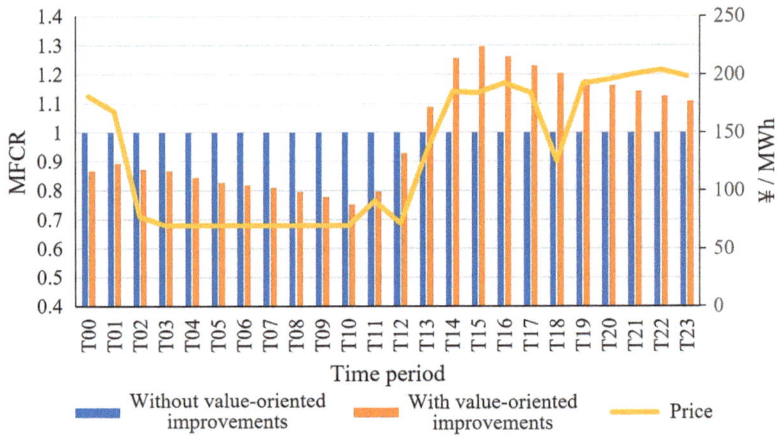

Fig. 3.7 MFCR comparison before and after the implementation of the value-oriented improvements in Sect. 3.2.2.3 (Aug 30), reprinted from [8], open access

3.3 System-Level Resource Aggregation

3.3.1 Problem Formulation

Let's consider a radial distribution sub-system with (1) a node set $\mathbf{N} = \{1, \ldots, \mathrm{N_N}\}$, (2) a generator set[1] $\mathbf{G} = \{1, \ldots, \mathrm{N_G}\}$, and 3) a branch set $\mathbf{L} = \{1, \ldots, \mathrm{N_L}\}$. Particularly, this chapter treats the branches as directed, i.e., the branch which connects nodes $i, j \in N$ is denoted by $i \longrightarrow j$ where node i is closer to the root node than node j. This chapter employs the widely-used Dist-Flow equations [14] to model the constraints of one distribution sub-system as follows:

(1) Active and reactive power injected into each node should equal their flow out of the node, i.e.,

$$P_{ij} - R_{ij} I_{ij} + \sum_{g \in \mathbf{G}} e_{jg} P_g + e_{j0} P_{\mathrm{PCC}} = \sum_{j \to k} P_{jk} + P_j, \forall j \in \mathbf{N}, \tag{3.16}$$

$$Q_{ij} - X_{ij} I_{ij} + \sum_{g \in \mathbf{G}} e_{jg} Q_g + e_{j0} Q_{\mathrm{PCC}} = \sum_{j \to k} Q_{jk} + Q_j, \forall j \in \mathbf{N}, \tag{3.17}$$

where P_{ij} and Q_{ij} are active and reactive branch power flows from node i to node j, respectively; I_{ij} is the squared current magnitude from node i to node j; P_g and Q_g are active and reactive generation levels of unit g, respectively; P_{PCC} and Q_{PCC} are active and reactive power flow exchanges at PCC, respectively; P_j and Q_j are active

[1] For illustration purposes, Sect. 3.3 suppose that the DERs aggregated from each node by the node-level resource aggregation method in Sect. 3.2 are in a generator shape, while its extension to any other convex shapes is straightforward.

and reactive power loads at node j, respectively; R_{ij} and X_{ij} are constant resistance and reactance from node i to node j; e_{jg} and e_{j0} are incident indicators.[2]

(2) Voltages change along each branch as

$$V_j = V_i - 2\left(R_{ij}P_{ij} + X_{ij}Q_{ij}\right) + \left(\left(R_{ij}\right)^2 + \left(X_{ij}\right)^2\right) I_{ij}, \forall i \to j, \tag{3.18}$$

where V_i is the squared voltage magnitude at node i.

(3) Voltages and currents affect branch flows by

$$V_i I_{ij} = \left(P_{ij}\right)^2 + \left(Q_{ij}\right)^2, \forall i \to j, \tag{3.19}$$

(4) The squared voltages should not exceed their upper bound $V_{\max,i}$ and lower bound $V_{\min,i}$, i.e.,

$$V_{\min,i} \le V_i \le V_{\max,i}, \forall i \in \mathbf{N}. \tag{3.20}$$

(5) The squared currents should not exceed their upper bound $I_{\max,ij}$ and lower bound $I_{\min,ij}$, i.e.,

$$I_{\min,ij} \le I_{ij} \le I_{\max,ij}, \forall i \to j, \tag{3.21}$$

(6) The generation levels should not exceed their upper and lower bounds, i.e.,

$$P_{\min,g} \le P_g \le P_{\max,g}, \forall g \in \mathbf{G}, \tag{3.22}$$

$$Q_{\min,g} \le Q_g \le Q_{\max,g}, \forall g \in \mathbf{G}, \tag{3.23}$$

where $P_{\min,g}$ and $P_{\max,g}$ are the lower and upper bounds of active generation levels of unit g; $Q_{\min,g}$ and $Q_{\max,g}$ are the lower and upper bounds of reactive generation levels of unit g.

(7) Power consumers can adjust their power demands between their upper bounds and lower bounds, i.e.,

$$P_{\min,n} \le P_n \le P_{\max,n}, \forall n \in \mathbf{N}, \tag{3.24}$$

$$Q_{\min,n} \le Q_n \le Q_{\max,n}, \forall n \in \mathbf{N}, \tag{3.25}$$

where $P_{\min,n}$ and $P_{\max,n}$ are the lower and upper bounds of active power loads at node n; $Q_{\min,n}$ and $Q_{\max,n}$ are the lower and upper bounds of reactive power loads at node n.

For ease of discussions, all the branch flows $\left(P_{ij}, Q_{ij}\right)$ are stacked as $(\mathbf{P}_{\mathrm{L}}, \mathbf{Q}_{\mathrm{L}})$ where $\mathbf{P}_{\mathrm{L}}$ and $\mathbf{Q}_{\mathrm{L}}$, all the generation levels $\left(P_g, Q_g\right)$ are stacked as $(\mathbf{P}_{\mathrm{G}}, \mathbf{Q}_{\mathrm{G}})$, all the power demands (P_n, Q_n) are stacked as $(\mathbf{P}_{\mathrm{N}}, \mathbf{Q}_{\mathrm{N}})$, all the squared voltages V_i are

[2] If unit g is connected to node j, e_{jg} equals one; otherwise, $e_{jg} = 0$. e_{j0} is defined in a similar manner.

stacked as $\mathbf{V}$, all the squared currents I_{ij} are stacked as $\mathbf{I}$, and the PCC power flow exchanges (P_{PCC}, Q_{PCC}) are stacked as $\mathbf{w}$.

In power industries, the variables in (3.16)–(3.25) can be categorized into (1) controllable variables $\mathbf{u} = \left[\mathbf{P}_{\mathrm{G}}^{\top}\ \mathbf{Q}_{\mathrm{G}}^{\top}\ \mathbf{P}_{\mathrm{N}}^{\top}\ \mathbf{Q}_{\mathrm{N}}^{\top}\right]^{\top}$ whose total number is $\mathrm{N_u} = 2\mathrm{N_G} + 2\mathrm{N_N}$, and (2) the state variables ($\mathbf{P}_{\mathrm{L}}, \mathbf{Q}_{\mathrm{L}}, \mathbf{V}, \mathbf{I}$) whose total number is $\mathrm{N_x} = 3\mathrm{N_L} + \mathrm{N_N}$. For ease of discussions, the constraints (3.16)–(3.25) are compacted as

$$\mathbf{f}\left(\mathbf{P}_{\mathrm{L}}, \mathbf{Q}_{\mathrm{L}}, \mathbf{V}, \mathbf{I}\right) = \mathbf{K}_1\mathbf{w} + \mathbf{K}_2\mathbf{u}, \tag{3.26}$$

$$\mathbf{V}_{\min} \le \mathbf{V} \le \mathbf{V}_{\max}, \tag{3.27}$$

$$\mathbf{I}_{\min} \le \mathbf{I} \le \mathbf{I}_{\max}, \tag{3.28}$$

$$\mathbf{u}_{\min} \le \mathbf{u} \le \mathbf{u}_{\max}, \tag{3.29}$$

where $\mathbf{f} = \left[\, f_1\ f_2 \ldots f_{\mathrm{N_e}} \,\right]^{\top}$ and $\mathrm{N_e} = 2\mathrm{N_L} + 2\mathrm{N_N}$. The equality Eq. (3.26) compact the power flow equations (3.16)–(3.19), the inequalities (3.27) and (3.28) compact the safety limits (3.20) and (3.21), the inequalities (3.29) compact the bounds (3.22)–(3.25) of controllable variables.

Based on the definition above, one permissible region of PCC power flow exchanges can be mathematically defined as

$$\mathbf{W} \triangleq \{\mathbf{w} \mid \exists\,(\mathbf{u}, \mathbf{P}_{\mathrm{L}}, \mathbf{Q}_{\mathrm{L}}, \mathbf{V}, \mathbf{I})\ \text{fulfilling (3.16)−(3.25)}\}. \tag{3.30}$$

Different from the transmission-level resource aggregation in Chap. 2, the multi-parametric programming theory cannot be applied to (3.16)–(3.25) to get the permissible region of PCC power flow exchanges due to the non-convex constraints (3.19). Consequently, this chapter will specially introduce another resource aggregation method to identify the permissible region of PCC power flow exchanges, as will be detailed in Sect. 3.3.2.

3.3.2 Fixed Point-Based Aggregation

3.3.2.1 Overview

The resource aggregation method for the permissible region of PCC power flow exchanges is outlined in Fig. 3.8. It iteratively explores inner sub-regions of the permissible region of PCC power flow exchanges. In each iteration, each of such inner sub-regions will be constructed around one current search point (i.e., the black dot in Fig. 3.8), which will be detailed in Sect. 3.3.2.2. Furthermore, an exploration strategy will be discussed to find more sub-regions based on the previously found sub-regions, as will be elaborated in Sect. 3.3.2.3. Particularly, the strategy will select the vertices of the found sub-regions as the candidates for the next search points (i.e., the

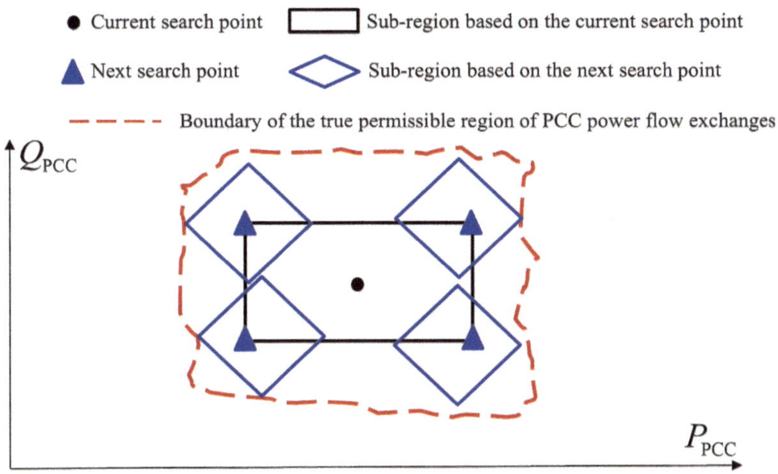

Fig. 3.8 Overview of the system-level resource aggregation method in Sect. 3.3.2, reprinted from [15], with permission from Elsevier

blue triangles in Fig. 3.8). Finally, the union of all the found sub-regions is employed as an approximation to the permissible region of PCC power flow exchanges.

3.3.2.2 Construction of an Inner Sub-Region

The resource aggregation method stems from a fixed-point representation of the power flow Eq. (3.26) when feasible PCC power flow exchanges $\mathbf{w}_0$ are given. Furthermore, Brouwer's fixed-point theorem from [16] is utilized to derive the conditions by which the solvability and safety of the fixed-point representation can be ensured. Finally, the derived conditions are used to formulate an inner sub-region of the permissible region of PCC power flow exchanges.

(1) Fixed-point representation of power flow equations
Suppose that there are $(\mathbf{u}_0, \mathbf{P}_{L0}, \mathbf{Q}_{L0}, \mathbf{V}_0, \mathbf{I}_0)$ with the given $\mathbf{w}_0$ to fulfill the constraints (3.26)–(3.29). For ease of discussions, (1) $(\mathbf{P}_L, \mathbf{Q}_L, \mathbf{V}, \mathbf{I})$ are stacked as $\mathbf{x}$, (2) $(\mathbf{P}_{L0}, \mathbf{Q}_{L0}, \mathbf{V}_0, \mathbf{I}_0)$ are denoted as $\mathbf{x}_0$, and (3) the ith element of $\mathbf{f}(\mathbf{x})$ in (3.26) is defined as $f_i(\mathbf{x})$. It is not hard to know that $f_i(\mathbf{x})$ is either linear or quadratic; thus, its second-order Taylor expansion is exact, i.e., $f_i(\mathbf{x})$ can be reformulated as

$$f_i(\mathbf{x}) = f_i(\mathbf{x}_0) + \nabla_{\mathbf{x}} f_i(\mathbf{x})|_{\mathbf{x}=\mathbf{x}_0} (\mathbf{x} - \mathbf{x}_0) + (\mathbf{x} - \mathbf{x}_0)^{\top} \frac{\nabla_{\mathbf{x}}^2 f_i(\mathbf{x})\big|_{\mathbf{x}=\mathbf{x}_0}}{2} (\mathbf{x} - \mathbf{x}_0), \tag{3.31}$$

where $\nabla_{\mathbf{x}} f_i(\mathbf{x})|_{\mathbf{x}=\mathbf{x}_0}$ and $\nabla_{\mathbf{x}}^2 f_i(\mathbf{x})\big|_{\mathbf{x}=\mathbf{x}_0}$ are the Jacobian and Hessian matrices of the function $f_i(\mathbf{x})$ at $\mathbf{x}_0$, respectively.

An equivalent formulation of $\mathbf{f}(\mathbf{x})$ in (3.26) can be got by stacking (3.31) for each $f_i(\mathbf{x})$, i.e., $\mathbf{f}(\mathbf{x})$ in (3.26) can be reformulated as

$$\mathbf{f}(\mathbf{x}) = \mathbf{f}(\mathbf{x}_0) + \nabla_{\mathbf{x}}\mathbf{f}(\mathbf{x})|_{\mathbf{x}=\mathbf{x}_0}(\mathbf{x} - \mathbf{x}_0) + \mathbf{R}, \tag{3.32}$$

where $\nabla_{\mathbf{x}}\mathbf{f}(\mathbf{x})|_{\mathbf{x}=\mathbf{x}_0}$ is assumed to be non-singular at $\mathbf{x}_0$; $\mathbf{R}$ is the second-order term and the ith element R_i in $\mathbf{R}$ is $(\mathbf{x} - \mathbf{x}_0)^\top \nabla^2_{\mathbf{x}} f_i(\mathbf{x})\big|_{\mathbf{x}=\mathbf{x}_0}(\mathbf{x} - \mathbf{x}_0)$.

Replacing $\mathbf{f}(\mathbf{x})$ in (3.26) with (3.32) to get

$$\mathbf{K}_1\mathbf{w} + \mathbf{K}_2\mathbf{u} = \mathbf{f}(\mathbf{x}_0) + \nabla_{\mathbf{x}}\mathbf{f}(\mathbf{x})|_{\mathbf{x}=\mathbf{x}_0}(\mathbf{x} - \mathbf{x}_0) + \mathbf{R}. \tag{3.33}$$

For $(\mathbf{x}_0, \mathbf{w}_{0,}, \mathbf{u}_0)$, the power flow Eq. (3.26) hold since $\mathbf{f}(\mathbf{x}_0) = \mathbf{K}_1\mathbf{w}_0 + \mathbf{K}_2\mathbf{u}_0$. This allows us to substitute $\mathbf{f}(\mathbf{x}_0) = \mathbf{K}_1\mathbf{w}_0 + \mathbf{K}_2\mathbf{u}_0$ into (3.33) to get

$$\begin{aligned} \mathbf{K}_1(\mathbf{w}_0 - \mathbf{w}_0) + \mathbf{K}_2(\mathbf{u} - \mathbf{u}_0) &= \nabla_{\mathbf{x}}\mathbf{f}(\mathbf{x})|_{\mathbf{x}=\mathbf{x}_0}(\mathbf{x} - \mathbf{x}_0) + \mathbf{R} \\ \Rightarrow \mathbf{K}_1\widetilde{\mathbf{w}} + \mathbf{K}_2\widetilde{\mathbf{u}} &= \nabla_{\mathbf{x}}\mathbf{f}(\mathbf{x})|_{\mathbf{x}=\mathbf{x}_0}\widetilde{\mathbf{x}} + \mathbf{R} \end{aligned}, \tag{3.34}$$

where $\widetilde{\mathbf{w}} = (\mathbf{w} - \mathbf{w}_0)$; $\widetilde{\mathbf{u}} = (\mathbf{u} - \mathbf{u}_0)$; $\widetilde{\mathbf{x}} = (\mathbf{x} - \mathbf{x}_0)$.

The Eq. (3.34) give the fixed-point representation of the power flow Eq. (3.26) as

$$\widetilde{\mathbf{x}} = \mathbf{F}_{\widetilde{\mathbf{w}}}\widetilde{\mathbf{w}} + \mathbf{F}_{\widetilde{\mathbf{u}}}\widetilde{\mathbf{u}} + \mathbf{F}_{\mathbf{x}}\mathbf{R}, \tag{3.35}$$

where $\mathbf{J} = \left(\nabla_{\mathbf{x}}\mathbf{f}(\mathbf{x})|_{\mathbf{x}=\mathbf{x}_0}\right)^\top \nabla_{\mathbf{x}}\mathbf{f}(\mathbf{x})|_{\mathbf{x}=\mathbf{x}_0}$; $\mathbf{F}_{\widetilde{\mathbf{u}}_{\mathrm{PCC}}} = \mathbf{J}^{-1}\left(\nabla_{\mathbf{x}}\mathbf{f}(\mathbf{x})|_{\mathbf{x}=\mathbf{x}_0}\right)^\top \mathbf{K}_1$; $\mathbf{F}_{\widetilde{\mathbf{u}}} = \mathbf{J}^{-1}\left(\nabla_{\mathbf{x}}\mathbf{f}(\mathbf{x})|_{\mathbf{x}=\mathbf{x}_0}\right)^\top \mathbf{K}_2$; $\mathbf{F}_{\mathbf{x}} = -\mathbf{J}^{-1}\left(\nabla_{\mathbf{x}}\mathbf{f}(\mathbf{x})|_{\mathbf{x}=\mathbf{x}_0}\right)^\top$.

The second-order term $\mathbf{R}$ in (3.35) is the second derivatives of the elements of $\mathbf{f}(\mathbf{x})$ in (3.26). From (3.26), the second-order terms associated with (3.16)–(3.18) are zero. Consequently, $\mathbf{R}$ can be reformulated as

$$\mathbf{R} = \left[\mathbf{0}^{(2\mathrm{N_N}+\mathrm{N_L})\times 1}\ \widetilde{\mathbf{R}}\right], \tag{3.36}$$

where $\widetilde{\mathbf{R}}$ stacks the second-order terms associated with (3.19):

$$\widetilde{\mathbf{R}} = -\widetilde{\mathbf{P}}_{\mathrm{L}} \otimes \widetilde{\mathbf{P}}_{\mathrm{L}} - \widetilde{\mathbf{Q}}_{\mathrm{L}} \otimes \widetilde{\mathbf{Q}}_{\mathrm{L}} + \left(\mathbf{e}_{\mathrm{VI}}\widetilde{\mathbf{V}}\right) \otimes \widetilde{\mathbf{I}}, \tag{3.37}$$

where $\mathbf{e}_{\mathrm{VI}}$ is the connection matrix between branches and nodes; $\otimes$ is the Hadamard product.

Based on the derivations above, the fixed-point representation in (3.35) can be equivalently given as

$$\widetilde{\mathbf{x}} = \mathbf{F}_{\widetilde{\mathbf{w}}}\widetilde{\mathbf{w}} + \mathbf{F}_{\widetilde{\mathbf{u}}}\widetilde{\mathbf{u}} + \mathbf{F}_{\widetilde{\mathbf{x}}}\widetilde{\mathbf{R}}. \tag{3.38}$$

(2) Conditions from Brouwer's fixed-point theorem

The core of this resource aggregation method is Brouwer's fixed-point theorem [16] in coming Theorem 3.1, to analyze the conditions by which the fixed-point representation (3.38) can be solved subject to the safety limits (3.27) and (3.28).

Theorem 3.1 *Suppose that there are (1) a compact and convex set* $\widetilde{\mathbf{X}}$*, and (2) a non-discrete mapping* $\mathbf{F}(\widetilde{\mathbf{x}})$*. If* $\mathbf{F}(\widetilde{\mathbf{x}}) \in \widetilde{\mathbf{X}}$ *for all* $\widetilde{\mathbf{x}} \in \widetilde{\mathbf{X}}$*, there is at least one point in* $\widetilde{\mathbf{X}}$ *such that* $\widetilde{\mathbf{x}} = \mathbf{F}(\widetilde{\mathbf{x}})$. ■

Brouwer's fixed-point theorem can be applied to (3.38) for its solvability by (1) regarding $(\widetilde{\mathbf{w}}, \widetilde{\mathbf{u}})$ as parameters, (2) $\widetilde{\mathbf{x}} = \left(\widetilde{\mathbf{P}}_{\mathrm{L}}, \widetilde{\mathbf{Q}}_{\mathrm{L}}, \widetilde{\mathbf{V}}, \widetilde{\mathbf{I}}\right)$ as variables, and 3) $\mathbf{F}(\widetilde{\mathbf{x}})$ as $\mathbf{F}_{\widetilde{\mathbf{w}}}\widetilde{\mathbf{w}} + \mathbf{F}_{\widetilde{\mathbf{u}}}\widetilde{\mathbf{u}} + \mathbf{F}_{\widetilde{\mathbf{x}}}\widetilde{\mathbf{R}}$. Particularly, the application conditions of Brouwer's fixed-point theorem rely on (1) the selection of a compact and convex set $\widetilde{\mathbf{X}}$, and (2) the guarantee of $\mathbf{F}(\widetilde{\mathbf{x}}) \in \widetilde{\mathbf{X}}$ for all $\widetilde{\mathbf{x}} \in \widetilde{\mathbf{X}}$. Consequently, let's next discuss how to fulfill these two conditions to identify an inner sub-region of the permissible region of PCC power flow exchanges.

Condition 1: Selection of $\widetilde{\mathbf{X}}$. If $\widetilde{\mathbf{X}}$ is selected as a subset of the constraints (3.27) and (3.28), the solutions from Brouwer's fixed-point theorem can ensure the fulfillment of the constraints (3.27) and (3.28). Motivated by this, a compact and convex set $\widetilde{\mathbf{X}}$ is selected as

$$\widetilde{\mathbf{X}} \triangleq \left\{ \left(\widetilde{\mathbf{P}}_{\mathrm{L}}, \widetilde{\mathbf{Q}}_{\mathrm{L}}, \widetilde{\mathbf{V}}, \widetilde{\mathbf{I}}\right) \middle| \begin{array}{l} \widetilde{\mathbf{V}}_{\min} \le \widetilde{\mathbf{V}} \le \widetilde{\mathbf{V}}_{\max} \\ \widetilde{\mathbf{I}}_{\min} \le \widetilde{\mathbf{I}} \le \widetilde{\mathbf{I}}_{\max} \end{array} \right\}, \tag{3.39}$$

where $\left(\widetilde{\mathbf{V}}_{\min}, \widetilde{\mathbf{V}}_{\max}, \widetilde{\mathbf{I}}_{\min}, \widetilde{\mathbf{I}}_{\max}\right)$ are the variables to be determined subject to the following bounds:

$$\begin{cases} \mathbf{V}_{\min} - \mathbf{V}_0 \le \widetilde{\mathbf{V}}_{\min} \le \widetilde{\mathbf{V}}_{\max} \le \mathbf{V}_{\max} - \mathbf{V}_0 \\ \mathbf{I}_{\min} - \mathbf{I}_0 \le \widetilde{\mathbf{I}}_{\min} \le \widetilde{\mathbf{I}}_{\max} \le \mathbf{I}_{\max} - \mathbf{I}_0 \end{cases}. \tag{3.40}$$

The determination of $\left(\widetilde{\mathbf{V}}_{\min}, \widetilde{\mathbf{V}}_{\max}, \widetilde{\mathbf{I}}_{\min}, \widetilde{\mathbf{I}}_{\max}\right)$ will be discussed later, since $\left(\widetilde{\mathbf{V}}_{\min}, \widetilde{\mathbf{V}}_{\max}, \widetilde{\mathbf{I}}_{\min}, \widetilde{\mathbf{I}}_{\max}\right)$ will affect how to guarantee $\mathbf{F}(\widetilde{\mathbf{x}}) \in \widetilde{\mathbf{X}}$ for all $\widetilde{\mathbf{x}} \in \widetilde{\mathbf{X}}$ in the coming Condition 2.

Condition 2: Guarantee of $\mathbf{F}(\widetilde{\mathbf{x}}) \in \widetilde{\mathbf{X}}$ for all $\widetilde{\mathbf{x}} \in \widetilde{\mathbf{X}}$. This condition can be fulfilled by selecting appropriate parameters $(\widetilde{\mathbf{w}}, \widetilde{\mathbf{u}})$, as explained below. Suppose that there are bounds $\widetilde{\mathbf{R}}_{\min}$ and $\widetilde{\mathbf{R}}_{\max}$ such that $\widetilde{\mathbf{R}}_{\min} \le \widetilde{\mathbf{R}} \le \widetilde{\mathbf{R}}_{\max}$ for all $\widetilde{\mathbf{x}} \in \widetilde{\mathbf{X}}$ (how to find $\widetilde{\mathbf{R}}_{\min}$ and $\widetilde{\mathbf{R}}_{\max}$ will be elaborated later). Let $\mathbf{A}_1$ be a selection matrix to make $\mathbf{A}_1 \left[\widetilde{\mathbf{P}}_{\mathrm{L}}^{\top}\ \widetilde{\mathbf{Q}}_{\mathrm{L}}^{\top}\ \widetilde{\mathbf{V}}^{\top}\ \widetilde{\mathbf{I}}^{\top}\right]^{\top} = \left[\widetilde{\mathbf{V}}^{\top}\ \widetilde{\mathbf{I}}^{\top}\right]^{\top}$. Define (1) $\mathbf{L}_{\widetilde{\mathbf{x}}} = \mathbf{A}_1\mathbf{F}_{\widetilde{\mathbf{x}}}$, and (2) a non-negative matrix $\mathbf{M}_{\widetilde{\mathbf{x}}}^{+} = \mathbf{L}_{\widetilde{\mathbf{x}}}^{+} - \mathbf{L}_{\widetilde{\mathbf{x}}}^{-}$ where $\mathbf{L}_{\widetilde{\mathbf{x}}}^{+}$ and $\mathbf{L}_{\widetilde{\mathbf{x}}}^{-}$ are the non-negative and non-positive parts of $\mathbf{L}_{\widetilde{\mathbf{x}}}$ with $\mathbf{L}_{\widetilde{\mathbf{x}}} = \mathbf{L}_{\widetilde{\mathbf{x}}}^{+} + \mathbf{L}_{\widetilde{\mathbf{x}}}^{-}$, respectively. The fulfillment of Condition 2 is claimed in the following Proposition 3.1 whose proof is given in Appendix 3.5.1.

Proposition 3.1 *Suppose that the following inequalities hold:*

$$\left[\widetilde{\mathbf{V}}_{\min}^{\top}\ \widetilde{\mathbf{I}}_{\min}^{\top}\right]^{\top} - \left[\widetilde{\mathbf{V}}_{\max}^{\top}\ \widetilde{\mathbf{I}}_{\max}^{\top}\right]^{\top} \le \mathbf{M}_{\widetilde{x}}^{+}\widetilde{\mathbf{R}}_{\min} - \mathbf{M}_{\widetilde{x}}^{+}\widetilde{\mathbf{R}}_{\max} \tag{3.41}$$

Given (3.41), there exists a vector $\mathbf{U}$ *such that* $\mathbf{U} + \mathbf{F}_{\widetilde{\mathbf{x}}}\widetilde{\mathbf{R}}$ *in* $\widetilde{\mathbf{X}}$ *for all* $\widetilde{\mathbf{x}} \in \widetilde{\mathbf{X}}$. *If* $\mathbf{U}$ *is formulated as* $\mathbf{F}_{\widetilde{\mathbf{w}}}\widetilde{\mathbf{w}} + \mathbf{F}_{\widetilde{\mathbf{u}}}\widetilde{\mathbf{u}}$, *then* $\mathbf{F}(\widetilde{\mathbf{x}}) \in \widetilde{\mathbf{X}}$ *for all* $\widetilde{\mathbf{x}} \in \widetilde{\mathbf{X}}$ *can be fulfilled.* ■

(3) Linear constraints to depict an inner sub-region

Proposition 3.1 inspires that linear constraints can be utilized to construct an inner sub-region of the permissible region of PCC power flow exchanges. As shown in

(3.63), once $\left(\widetilde{\mathbf{V}}_{\min}, \widetilde{\mathbf{V}}_{\max}, \widetilde{\mathbf{I}}_{\min}, \widetilde{\mathbf{I}}_{\max}\right)$ and $\left(\widetilde{\mathbf{R}}_{\min}, \widetilde{\mathbf{R}}_{\max}\right)$ that satisfy (3.41) are given, the following polytope $\widetilde{\mathbf{W}}$ in the domain of $(\widetilde{\mathbf{w}}, \widetilde{\mathbf{u}})$ can be given as

$$\widetilde{\mathbf{W}} \triangleq \left\{ (\widetilde{\mathbf{w}}, \widetilde{\mathbf{u}}) \middle| \begin{array}{l} \mathbf{b}_{\min} \le \mathbf{A}_1 \mathbf{F}_{\widetilde{\mathbf{w}}} \widetilde{\mathbf{w}} + \mathbf{A}_1 \mathbf{F}_{\widetilde{\mathbf{u}}} \widetilde{\mathbf{u}} \\ \mathbf{A}_1 \mathbf{F}_{\widetilde{\mathbf{w}}} \widetilde{\mathbf{w}} + \mathbf{A}_1 \mathbf{F}_{\widetilde{\mathbf{u}}} \widetilde{\mathbf{u}} \le \mathbf{b}_{\max} \end{array} \right\}, \tag{3.42}$$

where $\mathbf{b}_{\min} = \left[\widetilde{\mathbf{V}}_{\min}^{\top}\ \widetilde{\mathbf{I}}_{\min}^{\top}\right]^{\top} - \mathbf{L}_{\widetilde{\mathbf{x}}}^{-} \widetilde{\mathbf{R}}_{\max} - \mathbf{L}_{\widetilde{\mathbf{x}}}^{+} \widetilde{\mathbf{R}}_{\min}$ and $\mathbf{b}_{\max} = \left[\widetilde{\mathbf{V}}_{\max}^{\top}\ \widetilde{\mathbf{I}}_{\max}^{\top}\right]^{\top} - \mathbf{L}_{\widetilde{\mathbf{x}}}^{+} \widetilde{\mathbf{R}}_{\max} - \mathbf{L}_{\widetilde{\mathbf{x}}}^{-} \widetilde{\mathbf{R}}_{\min}$.

For any $(\widetilde{\mathbf{w}}, \widetilde{\mathbf{u}})$ within $\widetilde{\mathbf{W}}$, Brouwers's fixed-point theorem ensures the solvability of the fixed-point representation (3.38) of the power flow Eq. (3.26). Moreover, since $\widetilde{\mathbf{X}}$ in Brouwers's fixed-point theorem is selected as a subset of the safety limits (3.27) and (3.28), the solutions in (3.26) from Brouwer's fixed-point theorem fulfill (3.27) and (3.28) as well. Additionally, if the bounds $\mathbf{u}_{\min} - \mathbf{u}_0 \leqslant \widetilde{\mathbf{u}} \leqslant \mathbf{u}_{\max} - \mathbf{u}_0$ are added to $\widetilde{\mathbf{W}}$, the constraints (3.29) can be also satisfied. Consequently, $\widetilde{\mathbf{W}}$ intersecting with $\mathbf{u}_{\min} - \mathbf{u}_0 \leqslant \widetilde{\mathbf{u}} \leqslant \mathbf{u}_{\max} - \mathbf{u}_0$ guarantees the fulfillment of the constraints (3.16)–(3.25) in the domain of $(\widetilde{\mathbf{w}}, \widetilde{\mathbf{u}})$ in a linear form. Taking advantage of its linear form, the multi-parametric programming theory can be employed for its projection onto the space of $\widetilde{\mathbf{w}}$, resulting in an inner sub-region of the permissible region of PCC power flow exchanges. Although the sub-region is in the domain of $\widetilde{\mathbf{w}}$, it can be easily transformed to the domain of $\mathbf{w}$ by $\widetilde{\mathbf{w}} = (\mathbf{w} - \mathbf{w}_0)$.

Based on the rationale above, the key to constructing the linear constraints which formulate the inner sub-region is to determine $\left(\widetilde{\mathbf{V}}_{\min}, \widetilde{\mathbf{V}}_{\max}, \widetilde{\mathbf{I}}_{\min}, \widetilde{\mathbf{I}}_{\max}\right)$ and $\left(\widetilde{\mathbf{R}}_{\min}, \widetilde{\mathbf{R}}_{\max}\right)$ which satisfy the constraints (3.41). Let's recall that $\widetilde{\mathbf{R}}_{\min}$ and $\widetilde{\mathbf{R}}_{\max}$ are the lower and upper bounds of the second-order term $\widetilde{\mathbf{R}}$ when $(\mathbf{P}_{\mathrm{L}}, \mathbf{Q}_{\mathrm{L}}, \mathbf{V}, \mathbf{I})$ are over $\widetilde{\mathbf{X}}$ (3.39). It should be noted that $\widetilde{\mathbf{R}}$ contain three terms: two quadratic terms $-\widetilde{\mathbf{P}}_{\mathrm{L}} \otimes \widetilde{\mathbf{P}}_{\mathrm{L}}$ and $-\widetilde{\mathbf{Q}}_{\mathrm{L}} \otimes \widetilde{\mathbf{Q}}_{\mathrm{L}}$, and one bilinear term $\mathbf{e}_{\mathrm{VI}} \widetilde{\mathbf{V}} \otimes \widetilde{\mathbf{I}}$. Since $\left(\widetilde{\mathbf{P}}_{\mathrm{L}}, \widetilde{\mathbf{Q}}_{\mathrm{L}}, \widetilde{\mathbf{V}}, \widetilde{\mathbf{I}}\right)$ are decoupled over $\widetilde{\mathbf{X}}$, $\widetilde{\mathbf{R}}_{\max}$ and $\widetilde{\mathbf{R}}_{\min}$ can be also decomposed into the following three terms:

$$\widetilde{\mathbf{R}}_{\max} = \widetilde{\mathbf{R}}_{\max}^{(\mathrm{P})} + \widetilde{\mathbf{R}}_{\max}^{(\mathrm{Q})} + \widetilde{\mathbf{R}}_{\max}^{(\mathrm{VI})}, \tag{3.43}$$

$$\widetilde{\mathbf{R}}_{\min} = \widetilde{\mathbf{R}}_{\min}^{(\mathrm{P})} + \widetilde{\mathbf{R}}_{\min}^{(\mathrm{Q})} + \widetilde{\mathbf{R}}_{\min}^{(\mathrm{VI})}, \tag{3.44}$$

where $\left(\widetilde{\mathbf{R}}_{\max}^{(\mathrm{P})}, \widetilde{\mathbf{R}}_{\max}^{(\mathrm{Q})}, \widetilde{\mathbf{R}}_{\max}^{(\mathrm{VI})}, \widetilde{\mathbf{R}}_{\min}^{(\mathrm{P})}, \widetilde{\mathbf{R}}_{\min}^{(\mathrm{Q})}, \widetilde{\mathbf{R}}_{\min}^{(\mathrm{VI})}\right)$ can be formulated by considering their monotonicity over $\widetilde{\mathbf{X}}$, i.e.,

$$\begin{cases} \widetilde{\mathbf{R}}_{\max}^{(\mathrm{P})} = \mathbf{0}, \text{ if } \widetilde{\mathbf{P}}_{\mathrm{Lmin}} \le \mathbf{0} \le \widetilde{\mathbf{P}}_{\mathrm{Lmax}} \\ \widetilde{\mathbf{R}}_{\max}^{(\mathrm{P})} = -\widetilde{\mathbf{P}}_{\mathrm{Lmin}} \otimes \widetilde{\mathbf{P}}_{\mathrm{Lmin}}, \text{ if } \mathbf{0} \le \widetilde{\mathbf{P}}_{\mathrm{Lmin}} \le \widetilde{\mathbf{P}}_{\mathrm{Lmax}} \\ \widetilde{\mathbf{R}}_{\max}^{(\mathrm{P})} = -\widetilde{\mathbf{P}}_{\mathrm{Lmax}} \otimes \widetilde{\mathbf{P}}_{\mathrm{Lmax}}, \text{ if } \widetilde{\mathbf{P}}_{\mathrm{Lmin}} \le \widetilde{\mathbf{P}}_{\mathrm{Lmax}} \le \mathbf{0} \end{cases}, \tag{3.45}$$

$$\begin{cases} \widetilde{\mathbf{R}}_{\max}^{(\mathrm{Q})} = \mathbf{0}, \text{ if } \widetilde{\mathbf{Q}}_{\mathrm{Lmin}} \le \mathbf{0} \le \widetilde{\mathbf{Q}}_{\mathrm{Lmax}} \\ \widetilde{\mathbf{R}}_{\max}^{(\mathrm{Q})} = -\widetilde{\mathbf{Q}}_{\mathrm{Lmin}} \otimes \widetilde{\mathbf{Q}}_{\mathrm{Lmin}}, \text{ if } \mathbf{0} \le \widetilde{\mathbf{Q}}_{\mathrm{Lmin}} \le \widetilde{\mathbf{Q}}_{\mathrm{Lmax}} \\ \widetilde{\mathbf{R}}_{\max}^{(\mathrm{Q})} = -\widetilde{\mathbf{Q}}_{\mathrm{Lmax}} \otimes \widetilde{\mathbf{Q}}_{\mathrm{Lmax}}, \text{ if } \widetilde{\mathbf{Q}}_{\mathrm{Lmin}} \le \widetilde{\mathbf{Q}}_{\mathrm{Lmax}} \le \mathbf{0} \end{cases}, \tag{3.46}$$

$$\begin{cases} \widetilde{\mathbf{R}}_{\min}^{(P)} = \left\{-\widetilde{\mathbf{P}}_{\text{Lmin}} \otimes \widetilde{\mathbf{P}}_{\text{Lmin}},\ -\widetilde{\mathbf{P}}_{\text{L}\max} \otimes \widetilde{\mathbf{P}}_{\text{L}\max}\right\}_{\min}, \\ \text{if } \widetilde{\mathbf{P}}_{\text{Lmin}} \leq \mathbf{0} \leq \widetilde{\mathbf{P}}_{\text{L}\max} \\ \widetilde{\mathbf{R}}_{\min}^{(P)} = -\widetilde{\mathbf{P}}_{\text{L}\max} \otimes \widetilde{\mathbf{P}}_{\text{L}\max},\ \text{if } \mathbf{0} \leq \widetilde{\mathbf{P}}_{\text{Lmin}} \leq \widetilde{\mathbf{P}}_{\text{Lmax}} \\ \widetilde{\mathbf{R}}_{\min}^{(P)} = -\widetilde{\mathbf{P}}_{\text{Lmin}} \otimes \widetilde{\mathbf{P}}_{\text{Lmin}},\ \text{if } \widetilde{\mathbf{P}}_{\text{Lmin}} \leq \widetilde{\mathbf{P}}_{\text{L}\max} \leq \mathbf{0} \end{cases}, \tag{3.47}$$

$$\begin{cases} \widetilde{\mathbf{R}}_{\min}^{(Q)} = \left\{-\widetilde{\mathbf{Q}}_{\text{Lmin}} \otimes \widetilde{\mathbf{Q}}_{\text{Lmin}},\ -\widetilde{\mathbf{Q}}_{\text{Lmax}} \otimes \widetilde{\mathbf{Q}}_{\text{Lmax}}\right\}_{\min}, \\ \text{if } \widetilde{\mathbf{Q}}_{\text{Lmin}} \leq \mathbf{0} \leq \widetilde{\mathbf{Q}}_{\text{Lmax}} \\ \widetilde{\mathbf{R}}_{\min}^{(Q)} = -\widetilde{\mathbf{Q}}_{\text{Lmax}} \otimes \widetilde{\mathbf{Q}}_{\text{Lmax}},\ \text{if } \mathbf{0} \leq \widetilde{\mathbf{Q}}_{\text{Lmin}} \leq \widetilde{\mathbf{Q}}_{\text{Lmax}} \\ \widetilde{\mathbf{R}}_{\min}^{(Q)} = -\widetilde{\mathbf{Q}}_{\text{Lmin}} \otimes \widetilde{\mathbf{Q}}_{\text{Lmin}},\ \text{if } \widetilde{\mathbf{Q}}_{\text{Lmin}} \leq \widetilde{\mathbf{Q}}_{\text{Lmax}} \leq \mathbf{0} \end{cases}, \tag{3.48}$$

$$\widetilde{\mathbf{R}}_{\max}^{(VI)} = \left\{ \begin{matrix} \left(\mathbf{e}_{\text{VI}}\widetilde{\mathbf{V}}_{\min}\right) \otimes \widetilde{\mathbf{I}}_{\min},\ \left(\mathbf{e}_{\text{VI}}\widetilde{\mathbf{V}}_{\min}\right) \otimes \widetilde{\mathbf{I}}_{\max} \\ \left(\mathbf{e}_{\text{VI}}\widetilde{\mathbf{V}}_{\max}\right) \otimes \widetilde{\mathbf{I}}_{\min},\ \left(\mathbf{e}_{\text{VI}}\widetilde{\mathbf{V}}_{\max}\right) \otimes \widetilde{\mathbf{I}}_{\max} \end{matrix} \right\}_{\max}, \tag{3.49}$$

$$\widetilde{\mathbf{R}}_{\min}^{(VI)} = \left\{ \begin{matrix} \left(\mathbf{e}_{\text{VI}}\widetilde{\mathbf{V}}_{\min}\right) \otimes \widetilde{\mathbf{I}}_{\min},\ \left(\mathbf{e}_{\text{VI}}\widetilde{\mathbf{V}}_{\min}\right) \otimes \widetilde{\mathbf{I}}_{\max} \\ \left(\mathbf{e}_{\text{VI}}\widetilde{\mathbf{V}}_{\max}\right) \otimes \widetilde{\mathbf{I}}_{\min},\ \left(\mathbf{e}_{\text{VI}}\widetilde{\mathbf{V}}_{\max}\right) \otimes \widetilde{\mathbf{I}}_{\max} \end{matrix} \right\}_{\min}, \tag{3.50}$$

where $\widetilde{\mathbf{P}}_{\text{Lmin}}$ and $\widetilde{\mathbf{P}}_{\text{Lmax}}$ are the lower and upper bounds of $\widetilde{\mathbf{P}}_{\text{L}}$, respectively; $\widetilde{\mathbf{Q}}_{\text{Lmin}}$ and $\widetilde{\mathbf{Q}}_{\text{Lmax}}$ are the lower and upper bounds of $\widetilde{\mathbf{Q}}_{\text{L}}$, respectively; "$\{\mathbf{A}\}_{\max}$" (resp. "$\{\mathbf{A}\}_{\min}$") selects the maximum (resp. minimum) elements of each row of $\mathbf{A}$.

Substituting (3.43)–(3.50) into (3.42) returns $\widetilde{\mathbf{W}}$ in the domain of $(\widetilde{\mathbf{w}}, \widetilde{\mathbf{u}})$, by which to fulfill the inequalities (3.41) for the employment of Brouwer's fixed-point theorem. Particularly, this chapter will expand $\widetilde{\mathbf{W}}$ as large as possible. To achieve this purpose, let's first quantify how large the polytope $\widetilde{\mathbf{W}}$ is. Based on (3.42), $\widetilde{\mathbf{W}}$ is shaped by a series of parallel hyperplanes in the domain of $(\widetilde{\mathbf{w}}, \widetilde{\mathbf{u}})$. Based on this observation, the total distance between these hyperplanes across all the dimensions can be computed as one metric to reflect how large the polytope $\widetilde{\mathbf{W}}$ is, i.e., the following distance metric will be calculated:

$$-\mathbf{e}_1^{\top}\left\{\left\{\left[\widetilde{\mathbf{V}}_{\min}^{\top}\ \widetilde{\mathbf{I}}_{\min}^{\top}\right]^{\top} - \left[\widetilde{\mathbf{V}}_{\max}^{\top}\ \widetilde{\mathbf{I}}_{\max}^{\top}\right]^{\top} + \mathbf{M}_{\widetilde{\mathbf{x}}}^{+}\widetilde{\mathbf{R}}_{\max} - \mathbf{M}_{\widetilde{\mathbf{x}}}^{+}\widetilde{\mathbf{R}}_{\min}\right\} \cdot /\mathbf{H}\right\}, \tag{3.51}$$

where ./ is the component-wise division; the ith element of $\mathbf{H}$ is the root-sum-squares of the ith row of $\mathbf{L}_{\text{u}} = \left[\mathbf{A}_1\mathbf{F}_{\widetilde{\mathbf{w}}}\ \mathbf{A}_1\mathbf{F}_{\widetilde{\mathbf{u}}}\right]$.

Based on the rationale above, the following optimization problem can be designed to expand $\widetilde{\mathbf{W}}$ as large as possible:

$$\min \mathbf{e}_1^{\top}\left\{\left\{\begin{matrix} \left[\widetilde{\mathbf{V}}_{\min}^{\top}\ \widetilde{\mathbf{I}}_{\min}^{\top}\right]^{\top} - \left[\widetilde{\mathbf{V}}_{\max}^{\top}\ \widetilde{\mathbf{I}}_{\max}^{\top}\right]^{\top} \\ +\mathbf{M}_{\widetilde{\mathbf{x}}}^{+}\left(\widetilde{\mathbf{R}}_{\max}^{(P)} + \widetilde{\mathbf{R}}_{\max}^{(Q)} + \widetilde{\mathbf{R}}_{\max}^{(VI)}\right) \\ -\mathbf{M}_{\widetilde{\mathbf{x}}}^{+}\left(\widetilde{\mathbf{R}}_{\min}^{(P)} + \widetilde{\mathbf{R}}_{\min}^{(Q)} + \widetilde{\mathbf{R}}_{\min}^{(VI)}\right) \end{matrix}\right\} ./\mathbf{H}\right\}, \tag{3.52}$$

$$\text{s.t. Constraints (3.40)–(3.41) and (3.45)–(3.50),} \tag{3.53}$$

over $\widetilde{\mathbf{V}}_{\min}$, $\widetilde{\mathbf{V}}_{\max}$, $\widetilde{\mathbf{I}}_{\min}$, $\widetilde{\mathbf{I}}_{\max}$, $\widetilde{\mathbf{P}}_{\mathrm{Lmin}}$, $\widetilde{\mathbf{P}}_{\mathrm{Lmax}}$, $\widetilde{\mathbf{Q}}_{\mathrm{Lmin}}$, $\widetilde{\mathbf{Q}}_{\mathrm{Lmax}}$, $\widetilde{\mathbf{R}}^{(\mathrm{P})}_{\min}$, $\widetilde{\mathbf{R}}^{(\mathrm{P})}_{\max}$, $\widetilde{\mathbf{R}}^{(\mathrm{Q})}_{\min}$, $\widetilde{\mathbf{R}}^{(\mathrm{Q})}_{\max}$, $\widetilde{\mathbf{R}}^{(\mathrm{VI})}_{\min}$, $\widetilde{\mathbf{R}}^{(\mathrm{VI})}_{\max}$.

The operators "$\{\}_{\max}$" and "$\{\}_{\min}$" in the constraint (3.53) can be removed by additionally introducing binary variables [17]. In addition, the explicit formulation of $\left(\widetilde{\mathbf{R}}^{(\mathrm{P})}_{\max}, \widetilde{\mathbf{R}}^{(\mathrm{P})}_{\min}\right)$ requires the binary variables to indicate the rank among $\mathbf{0}$, $\widetilde{\mathbf{P}}_{\mathrm{Lmin}}$, and $\widetilde{\mathbf{P}}_{\mathrm{Lmax}}$. A similar situation occurs for the explicit formulation of $\left(\widetilde{\mathbf{R}}^{(\mathrm{Q})}_{\max}, \widetilde{\mathbf{R}}^{(\mathrm{Q})}_{\min}\right)$. These issues make the optimization problem (3.52) and (3.53) as a *mixed-integer nonlinear programming* (MINLP) problem which is hard to solve. To avoid binary variables, the following *nonlinear programming* (NLP) problem to solve part of the variables in the optimization problem (3.52) and (3.53) is designed:

$$\min \mathbf{e}_1^\top \left\{ \left\{ \left[\widetilde{\mathbf{V}}^\top_{\min}\ \widetilde{\mathbf{I}}^\top_{\min}\right]^\top - \left[\widetilde{\mathbf{V}}^\top_{\max}\ \widetilde{\mathbf{I}}^\top_{\max}\right]^\top + \mathbf{M}^+_{\tilde{\mathbf{x}}} \widetilde{\mathbf{R}}^{(\mathrm{VI})}_{\max} - \mathbf{M}^-_{\tilde{\mathbf{x}}} \widetilde{\mathbf{R}}^{(\mathrm{VI})}_{\min} \right\} ./\mathbf{H} \right\}, \tag{3.54}$$

$$\text{s.t.} \begin{cases} \mathbf{V}_{\min} - \mathbf{V}_0 \le \widetilde{\mathbf{V}}_{\min} \le \widetilde{\mathbf{V}}_{\max} \le \mathbf{V}_{\max} - \mathbf{V}_0 \\ \mathbf{I}_{\min} - \mathbf{I}_0 \le \widetilde{\mathbf{I}}_{\min} \le \widetilde{\mathbf{I}}_{\max} \le \mathbf{I}_{\max} - \mathbf{I}_0 \end{cases}, \tag{3.55}$$

$$\begin{cases} \widetilde{\mathbf{R}}^{(\mathrm{VI})}_{\min} \le \left(\mathbf{e}_{\mathrm{VI}} \widetilde{\mathbf{V}}_{\min}\right) \otimes \widetilde{\mathbf{I}}_{\min}, \widetilde{\mathbf{R}}^{(\mathrm{VI})}_{\min} \le \left(\mathbf{e}_{\mathrm{VI}} \widetilde{\mathbf{V}}_{\max}\right) \otimes \widetilde{\mathbf{I}}_{\min} \\ \widetilde{\mathbf{R}}^{(\mathrm{VI})}_{\min} \le \left(\mathbf{e}_{\mathrm{VI}} \widetilde{\mathbf{V}}_{\min}\right) \otimes \widetilde{\mathbf{I}}_{\max}, \widetilde{\mathbf{R}}^{(\mathrm{VI})}_{\min} \le \left(\mathbf{e}_{\mathrm{VI}} \widetilde{\mathbf{V}}_{\max}\right) \otimes \widetilde{\mathbf{I}}_{\max} \end{cases}, \tag{3.56}$$

$$\begin{cases} \widetilde{\mathbf{R}}^{(\mathrm{VI})}_{\max} \ge \left(\mathbf{e}_{\mathrm{VI}} \widetilde{\mathbf{V}}_{\min}\right) \otimes \widetilde{\mathbf{I}}_{\min}, \widetilde{\mathbf{R}}^{(\mathrm{VI})}_{\max} \ge \left(\mathbf{e}_{\mathrm{VI}} \widetilde{\mathbf{V}}_{\max}\right) \otimes \widetilde{\mathbf{I}}_{\min} \\ \widetilde{\mathbf{R}}^{(\mathrm{VI})}_{\max} \ge \left(\mathbf{e}_{\mathrm{VI}} \widetilde{\mathbf{V}}_{\min}\right) \otimes \widetilde{\mathbf{I}}_{\max}, \widetilde{\mathbf{R}}^{(\mathrm{VI})}_{\max} \ge \left(\mathbf{e}_{\mathrm{VI}} \widetilde{\mathbf{V}}_{\max}\right) \otimes \widetilde{\mathbf{I}}_{\max} \end{cases}, \tag{3.57}$$

over $\widetilde{\mathbf{V}}_{\min}$, $\widetilde{\mathbf{V}}_{\max}$, $\widetilde{\mathbf{I}}_{\min}$, $\widetilde{\mathbf{I}}_{\max}$, $\widetilde{\mathbf{R}}^{(\mathrm{VI})}_{\min}$, $\widetilde{\mathbf{R}}^{(\mathrm{VI})}_{\max}$. Let $\widetilde{\mathbf{x}}^*_{\mathrm{VI}} = (\widetilde{\mathbf{V}}^*_{\min}, \widetilde{\mathbf{V}}^*_{\max}, \widetilde{\mathbf{I}}^*_{\min}, \widetilde{\mathbf{I}}^*_{\max}, \widetilde{\mathbf{R}}^{(\mathrm{VI})*}_{\min}, \widetilde{\mathbf{R}}^{(\mathrm{VI})*}_{\max})$ denote the optimal solutions of the optimization problem (3.54)–(3.57).

Since the optimization problem (3.54)–(3.57) only solves a part of the variables in the optimization problem (3.52) and (3.53), the rest of the variables will be determined later using the following Proposition 3.2 whose proof is given in Appendix 3.5.2:

Proposition 3.2 *The variables of the optimization problem (3.52) and (3.53) can be divided into the following two parts:*

$$\begin{cases} \widetilde{\mathbf{x}}_{VI} = \left(\widetilde{\mathbf{V}}_{\min}, \widetilde{\mathbf{V}}_{\max}, \widetilde{\mathbf{I}}_{\min}, \widetilde{\mathbf{I}}_{\max}, \widetilde{\mathbf{R}}^{(VI)}_{\min}, \widetilde{\mathbf{R}}^{(VI)}_{\max}\right) \\ \widetilde{\mathbf{x}}_{Line} = \left(\widetilde{\mathbf{P}}_{\mathrm{Lmin}}, \widetilde{\mathbf{P}}_{Lmax}, \widetilde{\mathbf{Q}}_{Lmin}, \widetilde{\mathbf{Q}}_{Lmax}, \widetilde{\mathbf{R}}^{(P)}_{\min}, \widetilde{\mathbf{R}}^{(P)}_{\max}, \widetilde{\mathbf{R}}^{(Q)}_{\min}, \widetilde{\mathbf{R}}^{(Q)}_{\max}\right) \end{cases}. \tag{3.58}$$

If (1) $\widetilde{\mathbf{x}}_{\mathrm{VI}}$ in the optimization problem (3.52) and (3.53) equals $\widetilde{\mathbf{x}}^*_{\mathrm{VI}}$, and (2) $\widetilde{\mathbf{x}}_{\mathrm{Line}}$ in the optimization problem (3.52) and (3.53) equals zero, such solutions are feasible to the optimization problem (3.52) and (3.53). ■

Remark 3.1 Let's assign the reference directions of branches in such a way that branch $i \to j$ does not share its starting node i with other branches. In this context, $\widetilde{V}_i$ at node i can be indexed by its incident branch index $i \to j$. If $\widetilde{I}_{ij}$ is denoted as the $\tilde{l}$th element in $\widetilde{\mathbf{I}}$, the optimization problem (3.54)–(3.57) can be decoupled across branches as follows:

$$\min \left\{ \begin{array}{l} \left(\widetilde{V}_{\min,i} + \widetilde{I}_{\min,i} - \widetilde{V}_{\max,i} - \widetilde{I}_{\max,i}\right) \\ + \left(\mathrm{M}^{+}_{\widetilde{V},i} + \mathrm{M}^{+}_{\widetilde{I},i}\right)\left(\widetilde{R}^{(\mathrm{VI})}_{\max,i} - \widetilde{R}^{(\mathrm{VI})}_{\min,i}\right) \end{array} \right\} / H_i, \tag{3.59}$$

$$\text{s.t.} \begin{cases} V_{\min,i} - V_{0,i} \le \widetilde{V}_{\min,i} \le \widetilde{V}_{\max} \le V_{\max,i} - V_{0,i} \\ I_{\min,i} - I_{0,i} \le \widetilde{I}_{\min,i} \le \widetilde{I}_{\max} \le \widetilde{I}_{\max,i} - I_{0,i} \end{cases}, \tag{3.60}$$

$$\begin{cases} \widetilde{R}^{(\mathrm{VI})}_{\min,i} \le \widetilde{V}_{\min,i}\widetilde{I}_{\min,i},\ \widetilde{R}^{(\mathrm{VI})}_{\min,i} \le \widetilde{V}_{\max,i}\widetilde{I}_{\min,i} \\ \widetilde{R}^{(\mathrm{VI})}_{\min,i} \le \widetilde{V}_{\min,i}\widetilde{I}_{\max,i},\ \widetilde{R}^{(\mathrm{VI})}_{\min,i} \le \widetilde{V}_{\max,i}\widetilde{I}_{\max,i} \end{cases}, \tag{3.61}$$

$$\begin{cases} \widetilde{R}^{(\mathrm{VI})}_{\max,i} \ge \widetilde{V}_{\min,i}\widetilde{I}_{\min,i},\ \widetilde{R}^{(\mathrm{VI})}_{\max,i} \ge \widetilde{V}_{\max,i}\widetilde{I}_{\min,i} \\ \widetilde{R}^{(\mathrm{VI})}_{\max,i} \ge \widetilde{V}_{\min,i}\widetilde{I}_{\max,i},\ \widetilde{R}^{(\mathrm{VI})}_{\max,i} \ge \widetilde{V}_{\max,i}\widetilde{I}_{\max,i} \end{cases}, \tag{3.62}$$

over $\widetilde{V}_{\min,i}$, $\widetilde{V}_{\max,i}$, $\widetilde{I}_{\min,i}$, $\widetilde{I}_{\max,i}$, $\widetilde{R}^{(\mathrm{VI})}_{\min,i}$, $\widetilde{R}^{(\mathrm{VI})}_{\max,i}$. Particularly, $\mathrm{M}^{+}_{\widetilde{V},i}$ sum all the elements from the first $\mathrm{N_N}$ rows of the ith column of $\mathbf{M}^{+}_{\tilde{\mathbf{x}}}$, and $\mathbf{M}^{+}_{\widetilde{I},i}$ sums all the elements from the last $\mathrm{N_L}$ rows of the ith column of $\mathbf{M}^{+}_{\tilde{\mathbf{x}}}$.

After solving the optimization problem (3.59)–(3.62) for all the branches, the solutions of the optimization problem (3.54)–(3.57) can be obtained. Particularly, the minimum objective value of the optimization problem (3.59)–(3.62) is non-positive. This arises from the fact that zero is always feasible for the optimization problem (3.59)–(3.62). The decoupled treatment for the optimization problem (3.54)–(3.57) can be straightforwardly extended to the case where a branch $i \to j$ share its starting node i with other branches. ■

3.3.2.3 Exploration Strategy Based on Found Inner Sub-Regions

Since the sub-region found in Sect. 3.3.2.2 is only an inner estimation of the permissible region of PCC power flow exchanges, only a limited part of the whole permissible region may be covered by the found sub-region. Consequently, an exploration strategy is further discussed to find more sub-regions for a better estimation of the whole permissible region. The key idea is that a vertex of the found sub-region can be selected as a new search point to repeat the procedure in Sect. 3.3.2.2 if any of the following three conditions are fulfilled:

Condition 1. The vertex is only located on the boundary of one sub-region found in Sect. 3.3.2.2.

Condition 2. The number of the found sub-regions from Sect. 3.3.2.2 is smaller than a pre-set number.

Condition 3. The number of search points used in Sect. 3.3.2.2 does not exceed a pre-set number.

Condition 1 indicates that each vertex is expected to construct a new sub-region that has not been found. Conditions 2 and 3 allow the exploration strategy can stop at any time. When no new vertices satisfy any of the three conditions, the exploration strategy will terminate. Note that different starting points can be utilized to initialize the system-level aggregation technique. This provides an opportunity for a better estimation by getting more information from different locations in the true permissible region of PCC power flow exchanges. Such starting points can be selected based on the maximum and minimum PCC power transfers since they can provide certain information for limitations of PCC power flow exchanges.

3.3.3 Case Studies

The system-level resource aggregation method in Sect. 3.3.2 will be verified in the IEEE 33-bus and IEEE 136-bus test systems [18]. To examine the system-level resource aggregation method in Sect. 3.3.2, the permissible region of PCC power flow exchanges by brute-force search is first identified as the benchmark. Particularly, the benchmark comes from discretized PCC power flow exchanges. The range of $\mathbf{w} = (P_{\text{PCC}}, Q_{\text{PCC}})$ is evenly discretized as $101 \times 101 = 10201$ points. Given each point, its feasibility for the constraints (3.26)–(3.29) is checked so that the benchmark can be finalized.

In the IEEE 33-bus test system, the benchmark and the system-level resource aggregation method in Sect. 3.3.2 are compared in Fig. 3.9. The benchmark is marked using blue points. The system-level resource aggregation method in Sect. 3.3.2 provides four inner sub-regions to estimate the true permissible region of PCC power flow exchanges. Particularly, the cyan point marks the search point in the first iteration for constructing a sub-region, and the green points mark the search points in the second iteration. After two iterations, any of the three conditions in Sect. 3.3.2.3 is not fulfilled, and thus the system-level resource aggregation method in Sect. 3.3.2 stops. The union of the four sub-regions serves as the result of the system-level resource aggregation method in Sect. 3.3.2. Finally, 86.18% of the true permissible region is contained by the system-level resource aggregation method in Sect. 3.3.2.

The system-level resource aggregation method in Sect. 3.3.2 is further tested in the IEEE 136-bus test system. In this test system, the system-level resource aggregation method in Sect. 3.3.2 sets up four different points for its initialization. These starting points are selected as maximum and minimum PCC power transfers adhering to the constraints (3.26)–(3.29), which reflect the limitations of PCC power flow exchanges. 70 sub-regions are finally computed in the system-level resource aggregation method in Sect. 3.3.2. Their union is compared with the benchmark, as shown in Fig. 3.10. In Fig. 3.10, the blue solid lines are the boundaries of the benchmark, while each red shape indicates one sub-region from the system-level resource aggregation method in Sect. 3.3.2. Each sub-region of the system-level resource aggregation method in Sect. 3.3.2 may only cover a part of the benchmark, while the union of 70 sub-regions can cover 94.66% of the benchmark. This indicates that the exploration strategy in Sect. 3.3.2.3 paves a promising way for a good estimation of the benchmark by taking

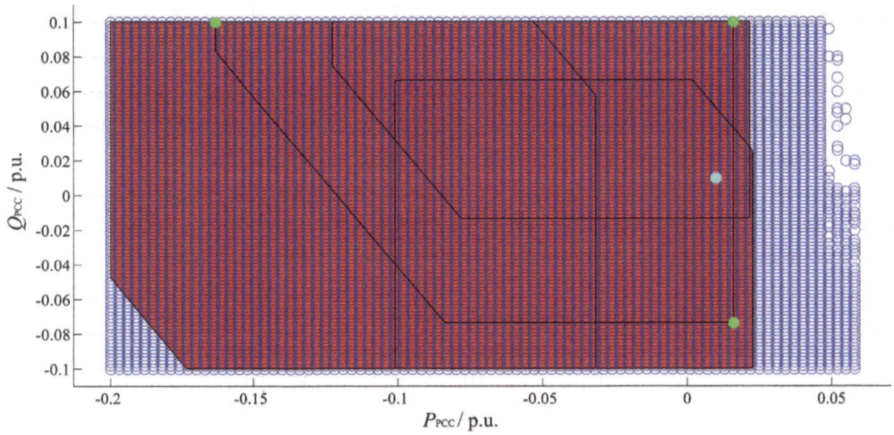

Fig. 3.9 Comparison of the system-level resource aggregation method in Sect. 3.3.2 with the benchmark in the IEEE 33-bus test system, reprinted from [15], with permission from Elsevier. The blue points are the benchmark, while the red shapes indicate sub-regions found by the system-level resource aggregation method in Sect. 3.3.2

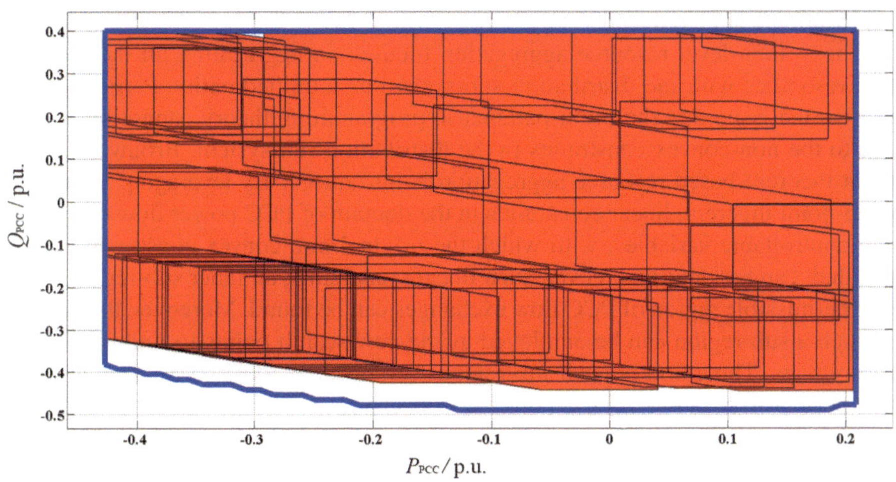

Fig. 3.10 Comparison of the system-level resource aggregation method in Sect. 3.3.2 with the benchmark in the IEEE 136-bus test system, reprinted from [15], with permission from Elsevier. The blue solid lines are the boundaries of the benchmark, while each red shape indicates a sub-region found by the system-level resource aggregation method in Sect. 3.3.2

the union of multiple sub-regions. Particularly, the union of any number of such sub-regions can serve as an inner estimation of the benchmark. This makes contributions to that the system-level resource aggregation method in Sect. 3.3.2 can guarantee the fulfillment of the constraints (3.26)–(3.29) when it stops at any time.

3.4 Summary

In this chapter, the resource aggregation for low-carbon power systems at the distribution level is discussed. The resource aggregation can be divided into the node-level resource aggregation and the system-level resource aggregation.

The node-level resource aggregation focuses on aggregating the DERs which are connected to the same node. It can be typically represented as a Minkowski sum problem while its computational complexity significantly increases with the number of individual DER constraints. To resolve this issue, a zonotope-based resource aggregation is introduced. Particularly, it is found that the feasible region of the individual DER constraints can be approximated by a zonotope and the Minkowski sum of such zonotope-based approximations can be explicitly formulated in a closed-form expression. This facilitates an efficiency improvement in the node-level resource aggregation. To further alleviate the approximation error due to the zonotope employment, it is found that the value of the node-level resource aggregation result can be employed. By taking the clearing price as the value weights in this chapter, the zonotope-based approximation can be reshaped for value-oriented improvements such that the impacts of approximation errors on the subsequent system-level resource aggregation can be reduced.

The system-level resource aggregation focuses on aggregating all the distribution resources with the distribution sub-system as a unified entity. Particularly, to resolve the issue that the multi-parametric programming theory cannot be applied due to the non-convex constraints in the distribution sub-system, Brouwer's fixed-point theorem-based resource aggregation is introduced. By leveraging Brouwer's fixed-point theorem, linear constraints in the domain of PCC power flow exchanges and controllable variables, with which the original non-convex constraints can be fulfilled, can be identified. Given such linear constraints, an inner sub-region of the true permissible region of PCC flow exchanges can be found. Moreover, the vertices of such a sub-region can be employed as the new search points to find more inner sub-regions. The union of such sub-regions can occupy a desirable coverage of the true permissible region of PCC power flow exchanges.

3.5 Appendixes

3.5.1 *Proof of Proposition 3.1*

The fulfillment of the inequalities (3.41) allows the existence of a a vector $\mathbf{U}$ such that

$$\begin{cases}\left[\widetilde{\mathbf{V}}_{\min}^{\top}\ \widetilde{\mathbf{I}}_{\min}^{\top}\right]^{\top}-\mathbf{L}_{\widetilde{\mathbf{x}}}^{-}\widetilde{\mathbf{R}}_{\max}-\mathbf{L}_{\widetilde{\mathbf{x}}}^{+}\widetilde{\mathbf{R}}_{\min}\leq\mathbf{A}_1\mathbf{U}\\ \mathbf{A}_1\mathbf{U}\leq\left[\widetilde{\mathbf{V}}_{\max}^{\top}\ \widetilde{\mathbf{I}}_{\max}^{\top}\right]^{\top}-\mathbf{L}_{\widetilde{\mathbf{x}}}^{+}\widetilde{\mathbf{R}}_{\max}-\mathbf{L}_{\widetilde{\mathbf{x}}}^{-}\widetilde{\mathbf{R}}_{\min}\end{cases}$$
$$\Rightarrow\begin{cases}\left[\widetilde{\mathbf{V}}_{\min}^{\top}\ \widetilde{\mathbf{I}}_{\min}^{\top}\right]^{\top}\leq\mathbf{A}_1\mathbf{U}+\mathbf{L}_{\widetilde{\mathbf{x}}}^{-}\widetilde{\mathbf{R}}_{\max}+\mathbf{L}_{\widetilde{\mathbf{x}}}^{+}\widetilde{\mathbf{R}}_{\min}\\ \mathbf{A}_1\mathbf{U}+\mathbf{L}_{\widetilde{\mathbf{x}}}^{+}\widetilde{\mathbf{R}}_{\max}+\mathbf{L}_{\widetilde{\mathbf{x}}}^{-}\widetilde{\mathbf{R}}_{\min}\leq\left[\widetilde{\mathbf{V}}_{\max}^{\top}\ \widetilde{\mathbf{I}}_{\max}^{\top}\right]^{\top}\end{cases}. \quad (3.63)$$

Since $\mathbf{L}_{\widetilde{\mathbf{x}}}^{+}$ is non-negative and $\mathbf{L}_{\widetilde{\mathbf{x}}}^{-}$ is non-positive, $\mathbf{L}_{\widetilde{\mathbf{x}}}\widetilde{\mathbf{R}}=\left(\mathbf{L}_{\widetilde{\mathbf{x}}}^{+}+\mathbf{L}_{\widetilde{\mathbf{x}}}^{-}\right)\widetilde{\mathbf{R}}$ can be bounded as

$$\mathbf{L}_{\widetilde{\mathbf{x}}}^{-}\widetilde{\mathbf{R}}_{\max}+\mathbf{L}_{\widetilde{\mathbf{x}}}^{+}\widetilde{\mathbf{R}}_{\min}\leq\mathbf{L}_{\widetilde{\mathbf{x}}}\widetilde{\mathbf{R}}\leq\mathbf{L}_{\widetilde{\mathbf{x}}}^{+}\widetilde{\mathbf{R}}_{\max}+\mathbf{L}_{\widetilde{\mathbf{x}}}^{-}\widetilde{\mathbf{R}}_{\min}. \quad (3.64)$$

Combining (3.63) and (3.64) yields $\mathbf{U}+\mathbf{F}_{\widetilde{\mathbf{x}}}\widetilde{\mathbf{R}}$ in $\widetilde{\mathbf{X}}$ for all $\widetilde{\mathbf{x}}\in\widetilde{\mathbf{X}}$, i.e., Proposition 3.1 holds.

3.5.2 Proof of Proposition 3.2

Let's consider an optimization problem by removing the constraints (3.41) from the optimization problem (3.52) and (3.53). This optimization problem can be decomposed as two optimization problems whose variables and constraints are respectively associated with $\widetilde{\mathbf{x}}_{\text{VI}}$ and $\widetilde{\mathbf{x}}_{\text{Line}}$ defined in (3.58), i.e., the following two optimization problems (3.65) and (3.66) and (3.67) and (3.68):

$$\min \mathbf{e}_1^{\top}\left\{\left\{\mathbf{M}_{\widetilde{\mathbf{x}}}^{+}\left(\left(\widetilde{\mathbf{R}}_{\max}^{(\mathrm{P})}-\widetilde{\mathbf{R}}_{\min}^{(\mathrm{P})}\right)+\left(\widetilde{\mathbf{R}}_{\max}^{(\mathrm{Q})}-\widetilde{\mathbf{R}}_{\min}^{(\mathrm{Q})}\right)\right)\right\}\cdot/\mathbf{H}\right\}, \quad (3.65)$$

$$\text{s.t. Constraints (3.45)–(3.48),} \quad (3.66)$$

over $\widetilde{\mathbf{P}}_{\text{Lmin}}, \widetilde{\mathbf{P}}_{\text{Lmax}}, \widetilde{\mathbf{Q}}_{\text{Lmin}}, \widetilde{\mathbf{Q}}_{\text{Lmax}}, \widetilde{\mathbf{R}}_{\min}^{(\mathrm{P})}, \widetilde{\mathbf{R}}_{\max}^{(\mathrm{P})}, \widetilde{\mathbf{R}}_{\min}^{(\mathrm{Q})}, \widetilde{\mathbf{R}}_{\max}^{(\mathrm{Q})}$.

$$\min \mathbf{e}_1^{\top}\left\{\left\{\begin{array}{l}\left[\widetilde{\mathbf{V}}_{\min}^{\top}\ \widetilde{\mathbf{I}}_{\min}^{\top}\right]^{\top}-\left[\widetilde{\mathbf{V}}_{\max}^{\top}\ \widetilde{\mathbf{I}}_{\max}^{\top}\right]^{\top}\\ +\mathbf{M}_{\widetilde{\mathbf{x}}}^{-}\widetilde{\mathbf{R}}_{\max}^{(\mathrm{VI})}-\mathbf{M}_{\widetilde{\mathbf{x}}}^{+}\widetilde{\mathbf{R}}_{\min}^{(\mathrm{VI})}\end{array}\right\}./\mathbf{H}\right\}, \quad (3.67)$$

$$\text{s.t. Constraints (3.40) and (3.55)–(3.57),} \quad (3.68)$$

over $\widetilde{\mathbf{V}}_{\min}, \widetilde{\mathbf{V}}_{\max}, \widetilde{\mathbf{I}}_{\min}, \widetilde{\mathbf{I}}_{\max}, \widetilde{\mathbf{R}}_{\min}^{(\mathrm{VI})}, \widetilde{\mathbf{R}}_{\max}^{(\mathrm{VI})}$.

For the optimization problem (3.65) and (3.66), $\left(\widetilde{\mathbf{R}}_{\max}^{(\mathrm{P})}-\widetilde{\mathbf{R}}_{\min}^{(\mathrm{P})}\right)$ and $\left(\widetilde{\mathbf{R}}_{\max}^{(\mathrm{Q})}-\widetilde{\mathbf{R}}_{\min}^{(\mathrm{Q})}\right)$ in the objective function (3.65) are non-negative. Consequently, the optimal objective value of the optimization problem (3.65) and (3.66) is zero, when $\widetilde{\mathbf{x}}_{\text{Line}}$ are taken as zero.

The optimization problem (3.67) and (3.68) is a reformulation of the optimization problem (3.54)–(3.57) after applying Lemma 3.1 in Appendix 3.5.3. This proof is further completed by showing that the constraints (3.41) can be fulfilled by the optimal solutions of the optimization problem (3.67) and (3.68) (i.e., the reformulation of the optimization problem (3.54)–(3.57)) and $\widetilde{\mathbf{x}}_{\text{Line}} = \mathbf{0}$. When $\widetilde{\mathbf{x}}_{\text{Line}}$ equal $\mathbf{0}$, the constraints (3.41) become

$$\left[\widetilde{\mathbf{V}}_{\min}^{\top}\,\widetilde{\mathbf{I}}_{\min}^{\top}\right]^{\top} - \left[\widetilde{\mathbf{V}}_{\max}^{\top}\,\widetilde{\mathbf{I}}_{\max}^{\top}\right]^{\top} \leq \mathbf{M}_{\widetilde{\mathbf{x}}}^{+}\widetilde{\mathbf{R}}_{\min}^{(\text{VI})} - \mathbf{M}_{\widetilde{\mathbf{x}}}^{+}\widetilde{\mathbf{R}}_{\max}^{(\text{VI})}. \tag{3.69}$$

Given the optimal solutions $\widetilde{\mathbf{x}}_{\text{VI}}^{*} = \left(\widetilde{\mathbf{V}}_{\min}^{*}, \widetilde{\mathbf{V}}_{\max}^{*}, \widetilde{\mathbf{I}}_{\min}^{*}, \widetilde{\mathbf{I}}_{\max}^{*}, \widetilde{\mathbf{R}}_{\min}^{(\text{VI})^{*}}, \widetilde{\mathbf{R}}_{\max}^{(\text{VI})*}\right)$ of the optimization problem (3.54)–(3.57), the fulfillment of the constraint (3.41) can be checked by solving the following LP problem:

$$\min_{\mathbf{s}_1} \mathbf{e}_1^{\top}\mathbf{s}_1 \tag{3.70}$$

$$\text{s.t.}\quad \mathbf{s}_1 \geq \mathbf{0}, \tag{3.71}$$

$$\begin{aligned}&\left[\left(\widetilde{\mathbf{V}}_{\min}^{*}\right)^{\top}\left(\widetilde{\mathbf{I}}_{\min}^{*}\right)^{\top}\right]^{\top} - \left[\left(\widetilde{\mathbf{V}}_{\max}^{*}\right)^{\top}\left(\widetilde{\mathbf{I}}_{\max}^{*}\right)^{\top}\right]^{\top} \\ &\leq -\mathbf{M}_{\widetilde{\mathbf{x}}}^{+}\widetilde{\mathbf{R}}_{\max}^{(\text{VI})^{*}} + \mathbf{M}_{\widetilde{\mathbf{x}}}^{+}\widetilde{\mathbf{R}}_{\min}^{(\text{VI})^{*}} + \mathbf{s}_1\end{aligned}. \tag{3.72}$$

Since the optimization problem (3.70)–(3.72) is an LP problem, the minimum value of the optimization problem (3.70)–(3.72) equals the maximum objective value of its following dual problem:

$$\max_{\boldsymbol{\lambda}} \boldsymbol{\lambda}^{\top}\left\{\begin{aligned}&\left[\left(\widetilde{\mathbf{V}}_{\min}^{*}\right)^{\top}\left(\widetilde{\mathbf{I}}_{\min}^{*}\right)^{\top}\right]^{\top} + \mathbf{M}_{\widetilde{\mathbf{x}}}^{+}\widetilde{\mathbf{R}}_{\max}^{(\text{VI})^{*}} \\ &-\left[\left(\widetilde{\mathbf{V}}_{\max}^{*}\right)^{\top}\left(\widetilde{\mathbf{I}}_{\max}^{*}\right)^{\top}\right]^{\top} - \mathbf{M}_{\widetilde{\mathbf{x}}}^{+}\widetilde{\mathbf{R}}_{\min}^{(\text{VI})^{*}}\end{aligned}\right\}, \tag{3.73}$$

$$\text{s.t.}\quad \mathbf{0} \leq \boldsymbol{\lambda} \leq \mathbf{1}. \tag{3.74}$$

The optimal solutions of the optimization problem (3.54)–(3.57) satisfy the constraints (3.69), if and only if the maximum objective value of the optimization problem (3.73) and (3.74) is zero. Consequently, the remaining part of this proof will show that the maximum objective value of the optimization problem (3.73) and (3.74) is zero.

The maximum objective value of the optimization problem (3.73) and (3.74) is a vertex of the constraints (3.74). i.e., it equals

$$\left\{z_0^{*}, z_1^{*}, \ldots, z_{\text{S}-1}^{*}\right\}_{\max}, \tag{3.75}$$

where $\text{S} = 2^{(\text{N}_\text{L}+\text{N}_\text{N})}$ is the vertex number and z_k^{*} denotes the value at the kth vertex as

$$\boldsymbol{\lambda}_k^\top \left\{ \begin{array}{l} \left[\left(\widetilde{\mathbf{V}}_{\min}^* \right)^\top \left(\widetilde{\mathbf{I}}_{\min}^* \right)^\top \right]^\top + \mathbf{M}_{\widetilde{\mathbf{x}}}^+ \widetilde{\mathbf{R}}_{\max}^{(\mathrm{VI})^*} \\ - \left[\left(\widetilde{\mathbf{V}}_{\max}^* \right)^\top \left(\widetilde{\mathbf{I}}_{\max}^* \right)^\top \right]^\top - \mathbf{M}_{\widetilde{\mathbf{x}}}^+ \widetilde{\mathbf{R}}_{\min}^{(\mathrm{VI})^*} \end{array} \right\}. \tag{3.76}$$

Particularly, all the elements in $\boldsymbol{\lambda}_0$ are zero and thus z_0^* equals zero. Meanwhile, at least one element in $\boldsymbol{\lambda}_k (k = 1, 2, \ldots, \mathrm{S} - 1)$ equals one. Based on Remark 3.1, (3.76) can be separated across branches, i.e.,

$$\begin{aligned} z_k^* &= \sum_{i=1}^{\mathrm{N_L}} z_{k,i}^* \\ &= \sum_{i=1}^{\mathrm{N_L}} \left\{ \begin{array}{l} \lambda_{k,i} \left(\widetilde{V}_{\min,i}^* - \widetilde{V}_{\max,i}^* \right) + \lambda_{k,i+\mathrm{N_N}} \left(\widetilde{I}_{\min,i}^* - \widetilde{I}_{\max,i}^* \right) \\ + \left(\lambda_{k,i} M_{\widetilde{V},i}^+ + \lambda_{k,i+\mathrm{N_N}} M_{\widetilde{I},i}^+ \right) \left(\widetilde{R}_{\max,i}^{(\mathrm{VI})^*} - \widetilde{R}_{\min,i}^{(\mathrm{VI})^*} \right) \end{array} \right\} \end{aligned}. \tag{3.77}$$

Lemma 3.2 in Appendix 3.5.3 implies that each $z_{k,i}^*$ in (3.77) is non-positive and thus z_k^* is non-positive. Since z_0^* is zero and z_k^* is non-positive, the maximum objective value of the optimization problem (3.73) and (3.74) is zero, i.e., the constraints (3.41) can be fulfilled under the optimal solutions of the optimization problem (3.54)–(3.57) and $\widetilde{\mathbf{x}}_{\mathrm{Line}} = \mathbf{0}$.

3.5.3 Proof of Two Lemmas

Let's consider (1) continuous real-valued variables $x_{\max}, x_{\min}, y_{\max}, y_{\min}, r_{\min}, r_{\min}$, (2) non-negative coefficients $\mathrm{M}_{\mathrm{x}}^+$ and $\mathrm{M}_{\mathrm{y}}^+$, and (3) real-valued bounds $x_{\max,0}, x_{\min,0}, y_{\max,0}, y_{\min,0}$ and $x_{\min,0} \leqslant 0 \leqslant x_{\max,0}$ and $y_{\min,0} \leqslant 0 \leqslant y_{\max,0}$. Define an optimization problem as

$$\min x_{\min} + y_{\min} - x_{\max} - y_{\max} + \left(\mathrm{M}_x^+ + \mathrm{M}_y^+ \right) \left(r_{\max} - r_{\min} \right), \tag{3.78}$$

$$\text{s.t.} \quad \left\{ \begin{array}{l} x_{\min,0} \leq x_{\min} \leq x_{\max} \leq x_{\max,0} \\ y_{\min,0} \leq y_{\min} \leq y_{\max} \leq y_{\max,0} \end{array} \right., \tag{3.79}$$

$$\left\{ \begin{array}{l} r_{\max} \geq x_{\min} y_{\min}, \; r_{\max} \geq x_{\max} y_{\max} \\ r_{\max} \geq x_{\max} y_{\min}, \; r_{\max} \geq x_{\min} y_{\max} \end{array} \right., \tag{3.80}$$

$$\left\{ \begin{array}{l} r_{\min} \leq x_{\max} y_{\min}, \; r_{\min} \leq x_{\min} y_{\max} \\ r_{\min} \leq x_{\max} y_{\max}, \; r_{\min} \leq x_{\min} y_{\min} \end{array} \right., \tag{3.81}$$

over $x_{\min}, x_{\max}, y_{\min}, y_{\max}, r_{\min}, r_{\max}$.

Lemma 3.1 *The optimal solutions $r_{\max}^*$ and $r_{\min}^*$ of the optimization problem (3.78)–(3.81) equal*

$$\begin{cases} r_{\max}^* = \left\{x_{\min}^* y_{\min}^*, x_{\min}^* y_{\max}^*, x_{\max}^* y_{\min}^*, x_{\max}^* y_{\max}^*\right\}_{\max} \\ r_{\min}^* = \left\{x_{\min}^* y_{\min}^*, x_{\min}^* y_{\max}^*, x_{\max}^* y_{\min}^*, x_{\max}^* y_{\max}^*\right\}_{\min} \end{cases}. \tag{3.82}$$

■

The proof of Lemma 3.1 is not provided in this chapter due to its straightforwardness.

Lemma 3.2 *Given* λ_x *and* $\lambda_y \in \{0, 1\}$ *and the optimal solutions (3.82), the following inequality holds*

$$\lambda_x \left(x_{\min}^* - x_{\max}^*\right) + \lambda_y \left(y_{\min}^* - y_{\max}^*\right) \leq \left(\lambda_x \mathrm{M}_x^+ + \lambda_y \mathrm{M}_y^+\right) \left(r_{\min}^* - r_{\max}^*\right). \tag{3.83}$$

■

Proof It should be noted that $\left(\lambda_x, \lambda_y\right)$ can be selected as (0, 0), (1, 1), (1, 0), and (0, 1). When $\left(\lambda_x, \lambda_y\right) = (0, 0)$ and $\left(\lambda_x, \lambda_y\right) = (1, 1)$, it is obviously that (3.83) holds. When $\left(\lambda_x, \lambda_y\right) = (1, 0)$, (3.83) becomes

$$\left(x_{\min}^* - x_{\max}^*\right) + \mathrm{M}_x^+ \left(r_{\max}^* - r_{\min}^*\right) \leq 0. \tag{3.84}$$

The optimization problem (3.78)–(3.81) can be solved by iteratively determining the variables in sequence. The variables $(y_{\min}, y_{\max})$ are first fixed and then variables $(x_{\min}, x_{\max}, r_{\min}, r_{\max})$ are determined. In this context, (3.84) is smaller than zero at optimum because $(x_{\min}, x_{\max}, r_{\min}, r_{\max})$ can attain zero and the optimization problem (3.78)–(3.81) is a minimization problem. The fulfillment of (3.83) can be similarly proven if $\left(\lambda_x, \lambda_y\right) = (0, 1)$. ■

References

1. Unlocking the potential of distributed energy resources: power system opportunities and best practices. https://www.iea.org/reports/unlocking-the-potential-of-distributed-energy-resources
2. Wang, J., Zhong, H., Wu, C., et al.: Incentivizing distributed energy resource aggregation in energy and capacity markets: an energy sharing scheme and mechanism design. Appl. Energy **252**, 113471 (2019)
3. Tsaousoglou, G., Pinson, P., Paterakis, N.G.: Transactive energy for flexible prosumers using algorithmic game theory. IEEE Trans. Sustain. Energy **12**(03), 1571–1581 (2021)
4. Pudjianto, D., Ramsay, C., Strbac, G.: Virtual power plant and system integration of distributed energy resources. IET Renew. Power Gener. **1**(1), 10–16 (2007)
5. Weibel, C.: Minkowski Sums of polytopes: combinatorics and computation, Ph.D. Thesis, Ecole Polytechnique Federale de Lausanne (2007)
6. Minkowski addition of convex polytopes. http://www.researchgate.net/publication/249915078_Minkowski_addition_of_convex_polytopes
7. Zhao, L., Hao, H., Zhang, W.: Extracting flexibility of heterogeneous deferrable loads via polytopic projection approximation. In: 2016 IEEE 55th IEEE Conference on Decision and Control (CDC) (2016)

8. Wang, Z.: Demand-side resources aggregation methods in electricity spot market, Master Thesis, Chongqing University (2021) (in Chinese)
9. Mueller, F.L., Sundström, O., Szabo, J., et al.: Aggregation of energetic flexibility using zonotopes. In: 2016 IEEE 54th IEEE Conference on Decision and Control (CDC) (2016)
10. Vandenberghe, L., Boyd, S., Wu, S.P.: Determinant maximization with linear matrix inequality constraints. SIAM J. Matrix Anal. Appl. **19**(2), 499–533 (1998)
11. Muller, F.L., Szabo, J., Sundstrom, O., et al.: Aggregation and disaggregation of energetic flexibility from distributed energy resources. IEEE Trans. Smart Grid **10**(2), 1205–1214 (2019)
12. Herre, L., Mathieu, J., Soder, L.: Impact of market timing on the profit of a risk-averse load aggregator. IEEE Trans. Power Syst. **35**(5), 3970–3980 (2020)
13. Li, Z., Li, T., Wu, W., et al.: Minkowski Sum based flexibility aggregating method of load dispatching for heat pumps. Autom. Electr. Power Syst. **45**(5), 14–21 (2019). (in Chinese)
14. Baran, M.E., Wu, F.F.: Optimal capacitor placement on radial distribution systems. IEEE Trans. Power Deliv. **4**(1), 725–734 (1989)
15. Lin, W., Zhao, C.: Improved characterization for AC-feasible power transfer regions of virtual power plants. Int. J. Electr. Power Energy Syst. **152**, 109260 (2023)
16. Spanier, E.H.: Algebraic Topology. Springer, Berlin (1994)
17. Tjeng, V., Xiao, K., Tedrake, R.: Evaluating robustness of neural networks with mixed integer programming. In: International Conference on Learning Representations (ICLR 2019) (2019)
18. VPP_AC. https://figshare.com/articles/dataset/VPP_AC/17082974

Chapter 4
Resource Aggregation for Low-Carbon Power Systems Interacted with Heterogeneous Energy Flows

4.1 Introduction

To address the global energy and environmental crises, the traditional energy system is transitioning towards a hybrid energy system. At the heart of this transition, power is integrated with gas and heat, fostering multi-energy coordination and complementarity. This facilitates the integration of renewable energy sources, in addition to supporting the low-carbon transformation of a modern energy system. Compared to conventional energy systems, the interaction of transmission and distribution subsystems with heterogeneous energy flows (i.e., gas flows and heating flows) can offer the following advantages [1, 2]:

- Enhanced multi-energy complementarity. Coupling equipment enables the co-generation and conversion of multiple energy flows, enhancing interdependence and flexibility among different energy sectors. This provides additional adjustable resources for more economical and secure operations.
- Inherent flexibility. The line pack of *gas systems* (GSs), the thermal inertia of *heating systems* (HSs), and loads offer natural flexibility to the transmission and distribution sub-systems. This flexibility promotes the integration of renewable energy and helps to smoothen load curves.
- Efficient co-generation. Co-generation equipment can utilize various types of energy in a cascading manner, improving energy quality and efficiency while reducing carbon emissions.

However, the interaction of transmission and distribution sub-systems with heterogeneous energy flows faces significant challenges due to the diversity of coupling modes and the heterogeneous properties of different energy flows. Additionally, the operational timescales of different energy systems vary widely. Mathematically, HSs and GSs are described by a set of high-dimensional, nonlinear *partial differential-algebraic equations* (PDAEs). These complexities—superimposed with different operators of GSs, HSs, transmission, and distribution sub-systems—pose substan-

W. Lin et al., *Coordinated Energy Management for Low-Carbon Power Systems*,
https://doi.org/10.1007/978-981-96-7533-3_4

tial obstacles to the coordination among GSs, HSs, transmission, and distribution sub-systems [3]. To address these issues, one potential solution is to aggregate these heterogeneous energy systems as permissible regions at the coupling points. After that, the operators of transmission and distribution sub-systems can directly perform the coordinated energy management by such permissible regions. This enables effective coordinated energy management across different energy systems.

This chapter focuses on resource aggregation methods for GSs and HSs to fully exploit energy conversion and co-generation within the system. The structure of this chapter is delineated below. Section 4.2 introduces the basic models of GSs and HSs, covering both static and dynamic models characterized by their individual timescales. Section 4.3 reveals a few analysis results for gas flows and heating flows due to their different timescales. The analysis results in Sect. 4.3 pave one promising way for resource aggregation of either one GS or HS even if they can simultaneously interact with transmissions and distribution sub-systems. Consequently, Sect. 4.4 introduces resource aggregation methods for individual energy systems, together with a *traverse-simulation-based concave hull* (TSCH) method for verification purposes. Section 4.5 provides the case studies, followed by the summary of this chapter in Sect. 4.6.

4.2 Modelling of Multi-energy Systems

In practical applications, the timescales of GSs and HSs typically range from tens of minutes to hours. In contrast, dynamic processes in the transmission and distribution sub-systems, such as electromagnetic and electromechanical transients, occur within microseconds to seconds, while static processes generally unfold over minutes. Consequently, a transmission or distribution sub-system primarily interacts with GSs and HSs on a timescale of minutes [4]. This motivates combined analysis predominantly on the minute-level timescale with the employment of static power flow models. On the other hand, the GS and HS models are categorized based on their respective timescales, leading to two types of energy flow analysis, i.e., static and dynamic energy flow analysis.

Static energy flow analysis focuses on determining the state distribution at a single time step. By solving sets of *algebraic equations* (AEs) derived from static GS and HS models, static energy flow analysis is commonly used for medium- to long-term analyses, such as planning or assessment problems, and for providing initial values for dynamic analysis. Dynamic energy flow analysis focuses on capturing the effects of energy flow dynamics and their interdependencies across multiple time steps. By solving PDAEs derived from dynamic GS and HS models, dynamic energy flow analysis is commonly used for real-time problems, such as state estimation, simulation, and economic dispatch.

4.2.1 Modelling of Gas Systems

4.2.1.1 Dynamic Model of Gas Systems

The transmission process of natural gas flow along the pipeline can be described by state and fluid dynamic equations. The state equations [3] are delineated by the gas pressure, the gas density, and temperature, i.e.,

$$p = \rho ZRT, \tag{4.1}$$

where p is the gas pressure; ρ is the gas density; Z is the gas compression factor; R is the gas constant; T is the temperature.

As for the fluid dynamic equations, they are usually built up based on the following assumptions for ease of analysis:

- The properties over the cross-section of the pipeline are assumed to be uniform, i.e., the gas flow is one-dimensional.
- The gas flow is isothermal since the temperature change due to thermal conduction is sufficiently slow to be neglected. It should be noted that, given the isothermal condition, the state equation for isothermal gas flow can be rewritten as

$$p = \rho c^2, \tag{4.2}$$

  where c is the sound speed.

Given the two assumptions, the fluid dynamic equations for isothermal gas flow contain the continuity and momentum equations. The continuity equation describes the mass conservation along the pipeline, i.e.,

$$\frac{\partial \rho}{\partial t} + \frac{\partial (\rho v)}{\partial x} = 0, \tag{4.3}$$

where v is the flow velocity. Particularly, the first and second terms in (4.3) reflect the mass change and flow increment of the natural gas, respectively.

The momentum equation describes the momentum conservation during flow transportation, i.e.,

$$\frac{\partial \left(\rho v^2\right)}{\partial x} + \frac{\partial (\rho v)}{\partial t} + \frac{\partial p}{\partial x} + \rho g \sin \sigma + \frac{\lambda_g \rho v^2}{2D} = 0, \tag{4.4}$$

where D is the pipeline diameter; g is the gravity acceleration; λ_g is the friction factor of a gas pipeline; σ is the pipeline inclination. The momentum Eq. (4.4) represents momentum conservation, where the four terms represent the momentum increment caused by convection, acceleration, surface stress, and friction resistance.

Moreover, the density and flow velocity are linked in the following equation:

$$q = Sv\rho, \tag{4.5}$$

where S is the pipeline cross section; q is the mass flow rate of natural gas. By replacing (4.5) into (4.3)–(4.4), the Eqs. (4.3)–(4.4) can be reformulated as

$$\begin{cases} \frac{\partial p}{\partial t} + \frac{c^2}{S}\frac{\partial q}{\partial x} = 0 \\ \frac{c^2}{S^2}\frac{\partial}{\partial t}\left(\frac{q^2}{p}\right) + \frac{\partial p}{\partial x} + \frac{1}{S}\frac{\partial q}{\partial t} + \frac{p}{c^2} g \sin\sigma + \frac{\lambda_g c^2 q^2}{2DS^2 p} = 0 \end{cases}. \tag{4.6}$$

Since (4.6) is too complex for analysis, two practical simplifications are usually fitted to the following two conditions: one is the zero value of the pipeline inclination σ, and the other one is that large amounts of literature have pointed out that the convection term in the momentum equation of gas flow dynamics can be neglected. Consequently, (4.6) can be reformulated as

$$\begin{cases} \frac{\partial p}{\partial t} + \frac{c^2}{S}\frac{\partial q}{\partial x} = 0 \\ \frac{\partial p}{\partial x} + \frac{1}{S}\frac{\partial q}{\partial t} + \frac{\lambda_g c^2 q|q|}{2DS^2 p} = 0 \end{cases}. \tag{4.7}$$

To avoid the nonlinearity in (4.7) for ease of analysis, the average flow velocity can be employed to linearize the model, thereby avoiding convergence and complexity issues while maintaining certain accuracy [3]. Particularly, the average flow velocity-based dynamic gas flow model can be expressed as

$$\begin{cases} \frac{\partial p}{\partial t} + \frac{c^2}{S}\frac{\partial q}{\partial x} = 0 \\ \frac{\partial p}{\partial x} + \frac{1}{S}\frac{\partial q}{\partial t} + \frac{\lambda_g w q}{2DS} = 0 \end{cases}, \tag{4.8}$$

where w is the average flow velocity of gas flow and can be formulated as

$$w\left(x, t_j\right) = \frac{1}{x_2 - x_1}\int_{x_1}^{x_2} v\left(x, t_j\right) dx, 0 \le x_1 \le x \le x_2 \le L, \tag{4.9}$$

$$w(x,t) = \frac{2\left(v_2^2 + v_1 v_2 + v_1^2\right)}{3\left(v_1 + v_2\right)}, x_1 \le x \le x_2, \tag{4.10}$$

where L is the pipeline length; x_1 and x_2 are the space variables, respectively; $v_1 = v\left(x_1, t\right)$ and $v_2 = v\left(x_2, t\right)$ are the flow velocity at x_1 and x_2, respectively.

Besides the distribution along the pipeline, gas states should also satisfy the conservation law at the junctions, i.e.,

$$p_b^{\circ} - p_{\mathrm{nd},k} = 0,\ p_n^{\mathrm{i}} - K_{\mathrm{cp},k} p_{\mathrm{nd},k} = 0,\ b \in \Phi_{\mathrm{g},k}^{\circ},\ n \in \Phi_{\mathrm{g},k}^{\mathrm{i}}, \tag{4.11}$$

$$\sum_b q_b^{\circ} - \sum_j q_j^{\mathrm{i}} - q_{\mathrm{nd},k} = 0,\ k \in \Theta_{\mathrm{g}},\ b \in \Phi_{\mathrm{g},k}^{\circ},\ j \in \Phi_{\mathrm{g},k}^{\mathrm{i}}, \tag{4.12}$$

where q_{nd} and p_{nd} are the node mass flow rate and the node pressure, respectively; $K_{\mathrm{cp},k}$ is the compression ratio if node k is a compressor, otherwise $K_{\mathrm{cp},k} = 1$; superscripts 'i' and 'o' are the symbols of inlet and outlet variables, respectively; $\Phi^{\mathrm{o}}_{\mathrm{g},k}$ and $\Phi^{\mathrm{i}}_{0,k}$ are the sets of pipelines ending and starting at node k in one GS, respectively; Θ_{g} is the node set in one GS.

Additionally, one GS should fulfill the following operational security constraints:

$$pn_i^{\min} \leq pn_i \leq pn_i^{\min}, i \in \Theta_{\mathrm{g,sr}}, \tag{4.13}$$

$$qn_i^{\min} \leq qn_i \leq qn_i^{\max}, i \in \Theta_{\mathrm{g,sr}}, \tag{4.14}$$

where $\Theta_{\mathrm{g,sr}}$ is the set of gas source nodes; $pn_i^{\min}$ and $pn_i^{\max}$ are the minimum and maximum node pressure in one GS, respectively; $qn_i^{\min}$ and $qn_i^{\max}$ are the minimum and maximum node mass flow rate of the gas sources, respectively.

4.2.1.2 Static Model of Gas Systems

At a larger timescale like hours or days, the gas flow dynamics turn to be static. Consequently, (4.8) can be rewritten into AEs by neglecting the differential terms regarding t. In this context, the first equation in (4.8) becomes

$$\frac{\mathrm{d}q}{\mathrm{d}x} = 0 \Rightarrow q = \text{constant}, \tag{4.15}$$

Substituting (4.15) into the second equation in (4.8) and neglecting the differential term of q regarding t, the following equations hold:

$$\frac{\mathrm{d}p}{\mathrm{d}x} + \frac{\lambda_{\mathrm{g}} c^2 q|q|}{2DS^2 p} = 0 \Rightarrow (p^{\mathrm{o}})^2 - \left(p^{\mathrm{i}}\right)^2 - K_{\mathrm{g}} q|q| = 0, \tag{4.16}$$

where p^{i} and p^{o} are the pipeline inlet and outlet pressures, respectively; K_{g} is the constant coefficient in the static GS model.

By substituting (4.16) into (4.11), all the pipeline pressures can be replaced with node pressures, i.e.,

$$p_{\mathrm{nd},k}^2 - p_{\mathrm{nd},n}^2 = K_{\mathrm{g},i} q_i \left|q_i\right|, i \in \Phi_{\mathrm{g}}, k \in \Theta_{\mathrm{g},i}^{+}, n \in \Theta_{\mathrm{g},i}^{-} \tag{4.17}$$

where $\Theta_{\mathrm{g},i}^{+}$ and $\Theta_{\mathrm{g},i}^{-}$ are the nodes at the inlet and outlet of pipeline i in one GS, respectively. Meanwhile, (4.12)–(4.14) remain unchanged in the static GS model. The variable distribution in one GS is shown in Fig. 4.1.

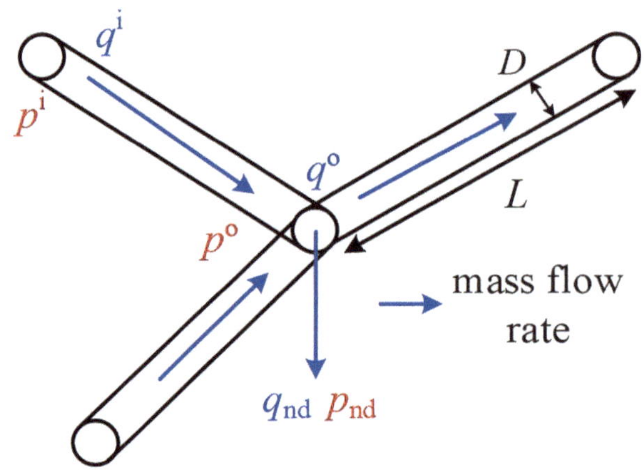

Fig. 4.1 Variable distribution in one GS. Reprinted from [9], with permission from IEEE

4.2.1.3 Gas-Fired Generators

One GS interacts with either transmission or distribution sub-systems through *gas-fired generators* (GTs), as illustrated in Fig. 4.2. The gas-fired generator comprises the compressor, *combustion chamber* (CC), and turbine [8].

In the GT, the compressor first consumes power to pressurize the incoming air. The high-pressure air is then sent to the CC which mixes and burns with the gas flow. Finally, its high-temperature gases drive the turbine to generate mechanical power. Note that the assumptions used to simplify the GT model include [8]:

- Constant physical-chemical properties. All the physical and chemical properties, such as specific heat capacity and adiabatic efficiency, are assumed to be constant.
- Negligible thermal losses. Thermal losses within the GT are ignored.
- No energy storage in compressor and turbine. The compressor and turbine are assumed to have no energy storage effect.

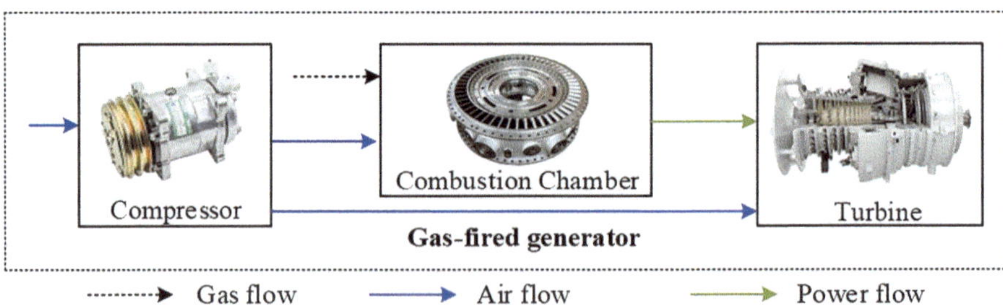

Fig. 4.2 GT-based energy interactions, reprinted from [10], with permission from IEEE

- Perfectly stirred combustion chamber. The CC is assumed to be a perfectly stirred region, ensuring thorough mixing and complete combustion of air and gas.

With the above assumptions, the temperature and pressure balance in the compressor can be expressed as

$$T_{\mathrm{C},t}^{\mathrm{o}} = T_{\mathrm{C},t}^{\mathrm{i}} \left[1 + \frac{1}{K_{\mathrm{C1}}} \left(K_{\mathrm{C2}}{}^{K_{\mathrm{C3}}} - 1 \right) \right], \tag{4.18}$$

$$p_{\mathrm{C},t}^{\mathrm{o}} = K_{\mathrm{C2}} p_{\mathrm{C},t}^{\mathrm{i}}, \tag{4.19}$$

where K_{C1} denotes the adiabatic efficiency; K_{C2} denotes the pressure ratio of the compressor; K_{C3} denotes the isentropic air coefficient; $T_{\mathrm{C},t}^{\mathrm{o}}$ and $T_{\mathrm{C},t}^{\mathrm{i}}$ are the outlet and inlet temperatures of the compressor, respectively; $p_{\mathrm{C},t}^{\mathrm{o}}$ and $p_{\mathrm{C},t}^{\mathrm{i}}$ are the outlet and inlet pressures of the compressor, respectively.

The power consumption of the compressor can be expressed as

$$P_{\mathrm{C},t} = m_{\mathrm{a},t} C_{\mathrm{a}} \left(T_{\mathrm{C},t}^{\mathrm{o}} - T_{\mathrm{C},t}^{\mathrm{i}} \right), \tag{4.20}$$

where m_{a} is the air mass flow rate; C_{a} is the specific heat capacity of the air; P_{C} is the power consumption of the compressor.

The overall thermal balance of the CC can be expressed as

$$\begin{aligned} K_{\mathrm{S1}} \frac{dT_{\mathrm{S},t}^{\mathrm{o}}}{dt} &= m_{\mathrm{a},t} C_{\mathrm{a}} \left(T_{\mathrm{C},t}^{\mathrm{o}} - 273.15 \right) + m_{\mathrm{g},t} \left(h_{\mathrm{g},t} + K_{\mathrm{S2}} \right) \\ &\quad - \left(m_{\mathrm{a},t} + m_{\mathrm{g},t} \right) C_{\mathrm{s}} \left(T_{\mathrm{S},t}^{\mathrm{o}} - 273.15 \right) \end{aligned}, \tag{4.21}$$

where h_{g} denotes the enthalpy of the gas flow; K_{S1} is the thermal storage coefficient of the combustion chamber; K_{S2} is the specific energy coefficient of the gas; C_{s} is the specific heat capacity of the smoke; $T_{\mathrm{S},t}^{\mathrm{o}}$ is the output temperature of the CC; m_{g} is the gas flow injecting into the CC. Particularly, the left-hand side of (4.21) denotes the temperature change due to the mixed combustion of gas and air. The terms of the right-hand side of (4.21) denote the power of input air, the power of input gas (including gas enthalpy and low-level heat energy), and the output power of the mixed smoke, respectively.

The inlet temperature and pressure of the turbine can be regarded as the temperature and pressure of the output smoke in the CC. The outlet temperature and output mechanical power of the turbine can be expressed as

$$T_{\mathrm{D},t}^{\mathrm{o}} = T_{\mathrm{S},t}^{\mathrm{o}} \left[1 - \left(1 - \left(\frac{p_{\mathrm{D},t}^{\mathrm{o}}}{p_{\mathrm{C},t}^{\mathrm{o}}} \right)^{K_{\mathrm{D1}}} \right) K_{\mathrm{D2}} \right], \tag{4.22}$$

$$P_{\mathrm{D},t} = \left(m_{\mathrm{a},t} + m_{\mathrm{g},t} \right) C_{\mathrm{s}} \left(T_{\mathrm{D},t}^{\mathrm{o}} - T_{\mathrm{s},t}^{\mathrm{o}} \right), \tag{4.23}$$

where K_{D1} and K_{D2} are the isentropic gas coefficient and internal efficiency of the turbine, respectively; $T^{o}_{D,t}$ is the outlet temperature of the turbine; $p^{o}_{D,t}$ is the outlet pressure of the turbine; $P_{D,t}$ is the mechanical output of the turbine.

The motion balance equation of rotors can be expressed as

$$\frac{dn}{dt} = \frac{K_{D3}\left(P_{D,t} - P_{C,t}\right) - P_{L,t}}{4\pi^2 Jn}, \tag{4.24}$$

where K_{D3} is the mechanical efficiency of the GT; n is the rotor speed; J is the moment of the inertia in the turbine. The transmission/distribution sub-system and GTs are coupled through the following equation:

$$K_{D3}\left(P_{D,t} - P_{C,t}\right) = P_{G,t}, i \in \Theta_e \cap \Theta_{gt}, \tag{4.25}$$

where Θ_{gt} is the set of gas-fired generators, $\Theta_e \cap \Theta_{gt}$ denotes the set of power buses connected with GTs. Equation (4.24) indicates that the output power of GTs equals the generated power at the generator bus.

For simplification purposes, a widely used GT model is expressed as a static algebraic equation, i.e.,

$$P_{gt} = \eta_{gt} q_{gt} h_g, \tag{4.26}$$

where η_{gt} is the efficiency of the GT; P_{gt} is the generated power of the GT; q_{gt} is the consumed gas flow of the GT.

4.2.2 *Modelling of Heating Systems*

One HS is a two-layer system composed of supply and return networks, where thermal power is transported by hot water flow. In one typical HS, the hot water is generated at thermal sources and transported by the supply pipelines. After being consumed by the thermal load, the water temperature drops and reflows into the return pipeline, thereby forming a flow cycle, as shown in Fig. 4.3. Based on the regulation mode, HS models can be categorized into the following two types:

- Quantity regulation mode. In this mode, operators adjust the mass flow rate to meet the thermal load requirements. This adjustment results in a coupled hydraulic and thermal model, reflecting the interdependence between water flow and temperature. In this condition, the HS model contains both hydraulic and thermal parts.
- Quality regulation mode. In this mode, operators adjust the water temperature while keeping the mass flow rate constant. Consequently, the HS model under quality regulation focuses solely on the thermal aspects, as the hydraulic conditions remain unchanged. In this condition, the HS model contains only thermal parts.

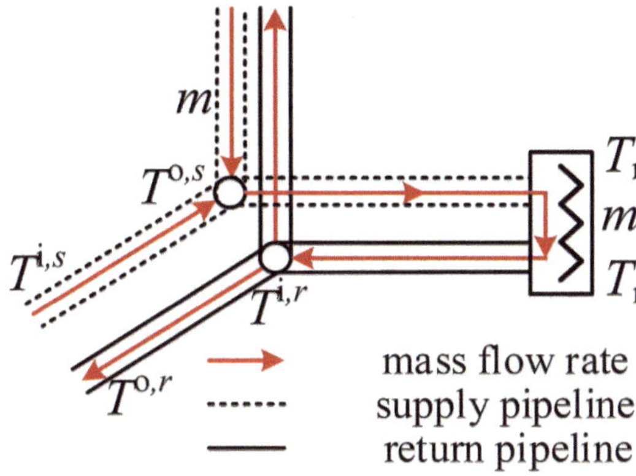

Fig. 4.3 Variable distribution in one HS. Reprinted from [11], with permission from IEEE

Due to the different characteristics of the two modes, the quantity regulation mode is mainly used in small-region systems for its higher efficiency, while the quality regulation mode is adopted in large-scale systems for its stable operation states [6].

4.2.2.1 Hydraulic Model of HSs

Under normal operations, the pressure and flow change over 1000 times faster than the temperature variation. The hydraulic dynamics are considered to be of less significance compared with the thermal dynamics. Consequently, the hydraulic model is built in a static form [5], containing a flow continuity Eq. (4.27), a loop pressure Eq. (4.28), and a pressure loss equation (4.29).

$$\mathbf{A}\mathbf{m} = \mathbf{m}_{\mathrm{nd}}, \tag{4.27}$$

$$\mathbf{B}\Delta\mathbf{p} = \mathbf{0}, \tag{4.28}$$

$$\Delta p_i = K_{\mathrm{f},i} m_i \left| m_i \right|, i \in \Phi_{\mathrm{h}}, \tag{4.29}$$

where $\mathbf{A}$ is the node-branch incidence matrix[1]; $\mathbf{m}$ is the vector of pipeline mass flow rate in one HS; $\mathbf{m}_{\mathrm{nd}}$ is the vector of node mass flow rate in one HS; $\mathbf{B}$ is the loop-branch incidence matrix[2]; K_{f} is the lumped pipeline resistance coefficient; $\Delta\mathbf{p}$ is the vector of pipeline pressure drop; Φ_h is the set of pipelines in the HS.

The operational security of the hydraulic parts in one HS can be expressed as

$$m_i^{\min} \le m_i \le m_i^{\max}, i \in \Phi_{\mathrm{h}}, \tag{4.30}$$

[1] If node i locates the inlet of pipeline, $j, a_{ij} = 1$. If node i locates the outlet of pipeline $j, a_{ij} = -1$. If it does not belong to the above two situations, $a_{ij} = 0$.

[2] If the loop i has the same direction as pipeline j, $b_{ij} = 1$. If the loop i has the reverse direction as pipeline j, $b_{ij} = -1$. If it does not belong to the above two situations,$b_{ij} = 0$.

where $m^{\min}$ and $m^{\max}$ are the minimum and maximum values of pipeline mass flow rate in one HS, respectively.

4.2.2.2 Thermal Model of HSs

Besides the hydraulic model, a thermal model is another part of one HS model. Firstly, the pipeline temperature should satisfy the energy conservation law. According to [7], the corresponding equation can be formulated by neglecting the fluid heat conduction and then expressed as

$$\frac{\partial T}{\partial t} + v\frac{\partial T}{\partial x} + \frac{v}{C_{\mathrm{w}} m \lambda_{\mathrm{h}}}(T - T_{\mathrm{a}}) = 0, \tag{4.31}$$

where T is the pipe temperature; C_{w} is the water specific heat capacity; λ_{h} is the pipeline thermal resistance. The three terms of the left-hand side of (4.31) represent the temperature variation of infinitesimal elements caused by the inner energy change, transverse heat input, and radial heat loss, respectively.

Commonly, the ambient temperature T_{a} in (4.31) is given as a set of constants, i.e., (4.31) can be reformulated as

$$\frac{\partial T'}{\partial t} + v\frac{\partial T'}{\partial x} + \frac{v}{C_{\mathrm{w}} m \lambda_{\mathrm{h}}} T' = 0, T' = T - T_{\mathrm{a}}, \tag{4.32}$$

where T' is the relative temperature that takes ambient temperature as a reference. For simplification, the superscript is neglected in the following contents and all the water temperature refers to the relative temperature if it is not specified, i.e. (4.31) is re-denoted as

$$\frac{\partial T}{\partial t} + v\frac{\partial T}{\partial x} + \frac{v}{C_{\mathrm{w}} m \lambda_{\mathrm{h}}} T = 0, \tag{4.33}$$

Secondly, the hot water flow mixing at the nodes should also satisfy the energy conservation law, thereby deriving the following temperature mixing equation:

$$T_{\mathrm{nd},k} \sum_{b \in \Phi^{\mathrm{i}}_{\mathrm{h},k}} m_b = \sum_{j \in \Phi^{\circ}_{\mathrm{h},k}} m_j T^{\circ}_j, k \in \Theta_h, \tag{4.34}$$

where Θ_{h} is the set of nodes in one HS; $\Phi^{\circ}_{\mathrm{h},k}$ and $\Phi^{\mathrm{i}}_{\mathrm{h},k}$ are the sets of pipelines ending and starting at node k in one HS; T_{nd} is the node temperature.

Thirdly, the node and pipeline inlet temperatures should satisfy the continuity equation, i.e.,

$$T^{\mathrm{i}}_j = T_{\mathrm{nd},k}, k \in \Theta_{\mathrm{h}}, j \in \Phi^{\mathrm{i}}_{\mathrm{h},k}. \tag{4.35}$$

Fourthly, the supply and return temperatures are linked by thermal power at nodes, as described in the following (4.36):

$$\phi_i = C_{\mathrm{w}} m_{\mathrm{nd},i} \left(T^{\mathrm{s}}_{\mathrm{nd},i} - T^{\mathrm{r}}_{\mathrm{nd},i} \right), k \in \Theta_{\mathrm{h}}, \tag{4.36}$$

where $T^{\mathrm{s}}_{\mathrm{nd}}$ and $T^{\mathrm{r}}_{\mathrm{nd}}$ are the node supply and return temperatures, respectively; ϕ is the thermal power.

Additionally, the operational security of the thermal parts in one HS is expressed as

$$T^{\mathrm{s}}_{\mathrm{nd},i,\min} \leq T^{\mathrm{s}}_{\mathrm{nd},i} \leq T^{\mathrm{s}}_{\mathrm{nd},i,\max}, i \in \Theta_h, \tag{4.37}$$

$$T^{\mathrm{r}}_{\mathrm{nd},i,\min} \leq T^{\mathrm{r}}_{\mathrm{nd},i} \leq T^{\mathrm{r}}_{\mathrm{nd},i,\max}, i \in \Theta_{\mathrm{h}}, \tag{4.38}$$

where $T^{\mathrm{s}}_{\mathrm{nd,min}}$ and $T^{s}_{\mathrm{nd,max}}$ are the minimum and maximum node supply temperatures, respectively; $T^{\mathrm{r}}_{\mathrm{nd,min}}$ and $T^{\mathrm{r}}_{\mathrm{nd,max}}$ are the minimum and maximum node return temperatures, respectively.

At a larger timescale like hours or days, the thermal dynamics similarly turn out to be static. In this condition, the differential term regarding t in (4.33) can be neglected, thereby yielding the following static HS model:

$$\frac{\mathrm{d}T}{\mathrm{d}x} + \frac{T}{C_{\mathrm{w}} m \lambda_{\mathrm{h}}} = 0 \Rightarrow T^{\mathrm{o}} = T^{\mathrm{i}} e^{\frac{-L}{C_{\mathrm{W}} m \lambda_{\mathrm{h}}}}, \tag{4.39}$$

where T^{i} and T^{o} are the pipeline inlet and outlet temperatures, respectively. Meanwhile, (4.34)–(4.36) in the dynamic HS model remains the same as the static HS model.

4.2.2.3 Coal-Fired Coupling Heating Power Units

One HS interacts with either transmission or distribution sub-systems through coal-fired *coupling heating power* (CHP) units, as illustrated in Fig. 4.4. The *gas-fired generators* (GTs) and the coal-fired *combined heat and power* (CHP) units. The coal-fired CHP unit is composed of several components, including the *coal-fired boiler* (CB), the *steam turbine* (ST), the reduction value, the *steam-water heat exchanger* (SHE), and the generator [12].

As illustrated in Fig. 4.4, the fuel flow injected into the CB is converted into a high-pressure steam flow with a specific temperature. A portion of this steam flow is directed to the ST, where it expands in the turbine cylinder. This drives the generator to produce power. The remaining steam flow from the CB is decompressed through a reduction valve and enters the SHE. From a physical perspective, the dynamic modelling of coal-fired CHP units should ideally encompass detailed models for each component. However, the dynamics within these components are governed by nonlinear *ordinary differential-algebraic equations* (ODAEs) and implicit physical-chemical equations, leading to challenges for their direct analysis. Consequently, this chapter simplifies the dynamics of coal-fired CHP units based on several practical assumptions and further introduces a simplified model for analysis. The assumptions are as follows:

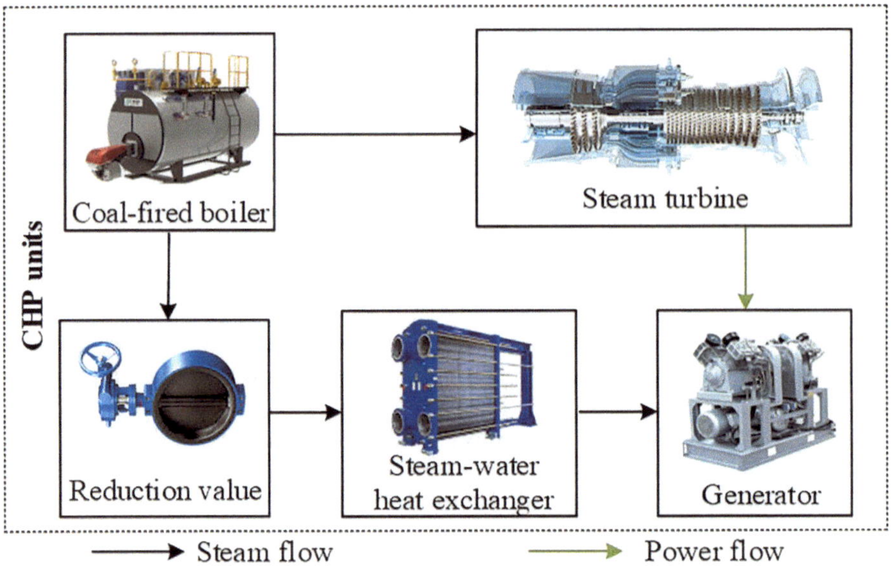

Fig. 4.4 Coal-fired CHP unit-based energy interactions, reprinted from [10], with permission from IEEE

- Boiler and superheater integration. The water wall and superheater inside the CB are considered as a whole, and an overall dynamic thermal balance model can be introduced for the CB.
- Steam drum stability. The water level in the steam drum inside the CB fluctuates slightly during regular operations. The thermal storage of the water/steam flow in the CB occupies only a small fraction of the overall thermal capacity, which is primarily determined by the metal walls. Consequently, the thermal storage coefficient of the CB can be treated as a constant.
- Inertia neglection. The rotational inertia of the generator's rotor and the thermal inertia of the ST's cylinder are ignored. An overall energy balance equation can be introduced to describe the dynamics in the ST and generator.
- Isenthalpic pressure reduction. The process of steam flow through the valve is considered an isenthalpic pressure reduction process.
- Phase transition ignorance: The phase transition process in the SHE is neglected, and an overall energy conservation equation can be established.
- Thermal loss neglection. Given the relatively higher thermal storage performance of the CB and ST, internal thermal losses in these components are ignored.
- Water temperature equivalence. The temperature of the input water flow is assumed to equal the return temperature at the sources in the HS, and the temperature of the output water flow is assumed to equal the supply temperature at the sources in the HS.

The dynamic model of coal-fired CHP units can be further established based on the assumptions above. For one CB, its input fuel flow can be described as

$$\phi_{B,t} = K_{B1} m_{B,t}, \tag{4.40}$$

where K_{B1} is the unit calorific value of the fuel; m_B is input fuel flow; ϕ_B is input power carried by the fuel flow.

The thermal balance of the CB can be expressed as:

$$K_{B3} \frac{dp_{B,t}}{dt} = -K_{T1} p_{T,t} - \phi_{H,t} + K_{B2} \phi_{B,t}, \tag{4.41}$$

where K_{B2} is the combustion efficiency of the CB; K_{B3} is the thermal storage coefficient; K_{T3} is the gain coefficient in the CB; $p_{B,t}$ is the pressure of the CB and ST; ϕ_H is the supplied thermal power for SHE. Particularly, the terms of the right-hand side of (4.41) denote the power sent into the ST, the thermal power used for heating, and the input power carried by the fuel, respectively.

The energy balance equation for the ST and generator can be expressed as

$$K_{T4} \frac{dP_{T,t}}{dt} = -P_{T,t} + K_{T2} K_{T3} p_{T,t}, \tag{4.42}$$

where K_{T2} is the opening of the intake valve in the ST; K_{T3} is the gain coefficients in the ST; K_{T4} is the time constant of the ST; P_T is the supplied power for the ST. Particularly, the terms of the left-hand side of (4.42) denote the generated power and the input power from the CB, respectively. According to [8], the pressures of the CB and the ST should satisfy the following equation:

$$p_{T,t} = p_{B,t} - K_{B4} K_{B2}^{1.3} \phi_{B,t}^{1.3}, \tag{4.43}$$

where K_{B4} is the resistance coefficient of the CB.

The energy balance equation for the SHE can be expressed as

$$\frac{dT_{H,t}^{o}}{dt} = \frac{\phi_{H,t}}{K_{H1}} - \frac{C_w m_H \left(T_{H,t}^{o} - T_{H,t}^{i}\right)}{K_{H1}} - \frac{K_{H2} T_{H,t}^{o}}{K_{H1}}, \tag{4.44}$$

where K_{H1} is the thermal storage coefficient of the SHE; K_{H2} is the thermal conductivity coefficient of SHE; m_H is the input water flow rate of the SHE; $T_{H,t}^{o}$ is the output temperature of the SHE; $T_{H,t}^{i}$ is the input temperature of the SHE.

To deal with the implicit ODAEs in (4.41), (4.42) and (4.44) for ease of further analysis, this chapter adopts the backward Euler scheme to discretize (4.41), (4.42) and (4.44). Then, the above equations become

$$p_{B,t} = \frac{\Delta t \left(K_{B2} K_{B1} m_{B,t} - K_{T1} p_{T,t} - \phi_{H,t}\right)}{K_{B3}} + p_{B,t-1}, \tag{4.45}$$

$$P_{\mathrm{T},t} = \frac{K_{\mathrm{T4}} P_{\mathrm{T},t-1} + \Delta t K_{\mathrm{T2}} K_{\mathrm{T3}} p_{\mathrm{T},t}}{K_{\mathrm{T4}} + \Delta t}, \tag{4.46}$$

$$T^{\mathrm{o}}_{\mathrm{H},t} = \frac{K_{\mathrm{H1}} T^{\mathrm{o}}_{\mathrm{H},t-1} + \Delta t \phi_{\mathrm{H},t} + \Delta t C_{\mathrm{w}} m^{\mathrm{i}}_{\mathrm{H},T_{\mathrm{H},t}}}{(K_{\mathrm{H1}} + \Delta t C_{\mathrm{w}} m_{\mathrm{H}} + \Delta t K_{\mathrm{H2}})}, \tag{4.47}$$

where Δt is the discrete time step. Meanwhile, the CHP unit should satisfy the following operational security constraints:

$$m_{\mathrm{B,min}} \le m_{\mathrm{B},t} \le m_{\mathrm{B,max}}, \tag{4.48}$$

$$p_{\mathrm{B,min}} \le p_{\mathrm{B},t} \le p_{\mathrm{B,max}}, \tag{4.49}$$

$$p_{\mathrm{T,min}} \le p_{\mathrm{T},t} \le p_{\mathrm{T,max}}, \tag{4.50}$$

$$\phi_{\mathrm{H,min}} \le \phi_{\mathrm{H},t} \le \phi_{\mathrm{H,max}}, \tag{4.51}$$

$$P_{\mathrm{B,min}} \le P_{\mathrm{B},t} \le P_{\mathrm{B,max}}, \tag{4.52}$$

$$P_{\mathrm{T},t} + \phi_{\mathrm{H},t} \le K_{\mathrm{B2}} \phi_{\mathrm{B},t}, \tag{4.53}$$

where $m_{\mathrm{B,min}}$ and $m_{\mathrm{B,max}}$ denote the minimum and maximum input fuel flow of the CB, respectively; $p_{\mathrm{B,min}}$ and $p_{\mathrm{B,max}}$ denote the minimum and maximum pressures of the CB, respectively; $p_{\mathrm{T,min}}$ and $p_{\mathrm{T,max}}$ denote the minimum and maximum pressures of the ST, respectively; $\phi_{\mathrm{H,min}}$ and $\phi_{\mathrm{H,max}}$ denote the minimum and maximum thermal power output of the SHE, respectively; $P_{\mathrm{T,min}}$ and $P_{\mathrm{T,max}}$ denote the minimum and maximum power outputs of the ST, respectively. Equation (4.53) indicates that the total output of the CHP unit should not exceed the input power.

According to the topological connections shown between the transmission/distribution sub-system and HS coupled by the coal-fired CHP units, the electric and thermal states are coupled through the following equations:

$$P_{\mathrm{T},t} = P_{\mathrm{G}i,t}, i \in \Theta_{\mathrm{e}} \cap \Theta_{\mathrm{chp}}, \tag{4.54}$$

$$m_i = -m_{\mathrm{H}}, i \in \Theta_{\mathrm{h,sr}} \cap \Theta_{\mathrm{chp}}, \tag{4.55}$$

$$T^{\mathrm{o}}_{\mathrm{H},t} = T^{\mathrm{s}}_{i,t}, T^{\mathrm{i}}_{\mathrm{H},s} = T^{\mathrm{r}}_{i,t}, i \in \Theta_{\mathrm{h,sr}} \cap \Theta_{\mathrm{chp}}, \tag{4.56}$$

where Θ_{chp} is the set of CHP units, $\Theta_{\mathrm{e}} \cap \Theta_{\mathrm{chp}}$ denotes the set of electric buses connected with CHP units; $\Theta_{\mathrm{h,\,sr}}$ is the set of source nodes in the HS, $\Theta_{\mathrm{h,sr}} \cap \Theta_{\mathrm{chp}}$ denotes the set of thermal source nodes connected with CHP units. In the above equations, (4.54) indicates that the power output of the ST in the CHP unit equals the generated power at the generator bus; (4.55) indicates that the input water flow rate of the SHE in the CHP unit equals the nodal mass flow rate at the source nodes in the HS; (4.56) indicates that the input and output temperatures of the SHE in the CHP unit equal the return and supply temperatures at the source nodes in the HS.

Since the equations above are complex for analysis, a widely used CHP unit model is expressed as a set of static algebraic equations, as shown below [13].

$$\begin{cases} 0 \leq P_{\text{chp},t} \leq P_{\text{chp}}^{\max} \\ 0 \leq \phi_{\text{chp},t} \leq \phi_{\text{chp}}^{\max} \\ k_{\text{chp1}}\phi_{\text{chp},t} \leq P_{\text{chp},t} \leq P_{\text{chp}}^{\max} - k_{\text{chp2}}\phi_{\text{chp},t} \end{cases}, \tag{4.57}$$

where k_{chp1} and k_{chp2} are the electric-thermal ratios in the CHP units, respectively; P_{chp1} and ϕ_{chp} are the power output and thermal output of the CHP units, respectively; $P_{\text{chp}}^{\max}$ is the maximum output of the CHP units; $\phi_{\text{chp}}^{\max}$ is the maximum thermal output of the CHP units.

4.3 Analysis for Heterogeneous Energy Flows

The slow dynamics of GSs and HSs mean that the distributions of gas and thermal states are influenced not only by the previous operational boundaries but also by the initial state distribution. This fact affects the permissible region of one GS or HS or their combination, making the permissible region characterization dependent on both the current and historical states. The integration of power, gas, and thermal parts means that the thermal and gas flow dynamics influence the interconnected transmission/distribution sub-system through coupling units, causing the transmission/distribution sub-system to be affected by the current and historical states within GSs and HSs. As a result, the permissible region of one transmission/distribution sub-system at each time step is no longer static and exhibits time-varying properties.

Let's consider the system illustrated in Fig. 4.5 as an example, where the power and thermal loads are fixed, and $\{y_1, y_2, y_3\}$ represents a set of secure generator power injections at time $t = t_0$. As the temperature component, determined by the initial states, gradually decreases over time, the CHP units must generate more thermal power to maintain the temperature balance in the HS. Given the fixed power capacity of the CHP units, the increasing thermal power output reduces the available range for electric power generation. Consequently, the permissible region shrinks at $t = t_1$ and $t = t_2$, making $\{y_2, y_3\}$ progressively insecure. To ensure the operational feasibility of the IES, it is essential to identify secure operational points that account for thermal dynamics throughout the sequential process (Fig. 4.6).

Based on the analysis above, the permissible regions of GSs and HSs differ significantly from those of transmission and distribution sub-systems in the following perspectives:

- The operational security at the current moment is determined jointly by states at the current moment and the state at historical moments.

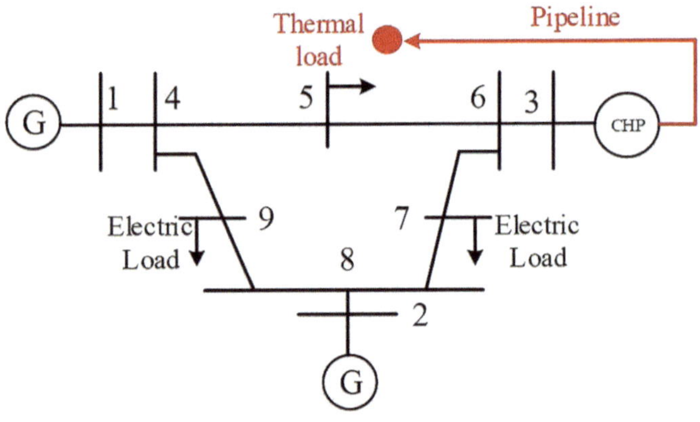

Fig. 4.5 Structures of the illustrative case, reprinted from [10], with permission from IEEE

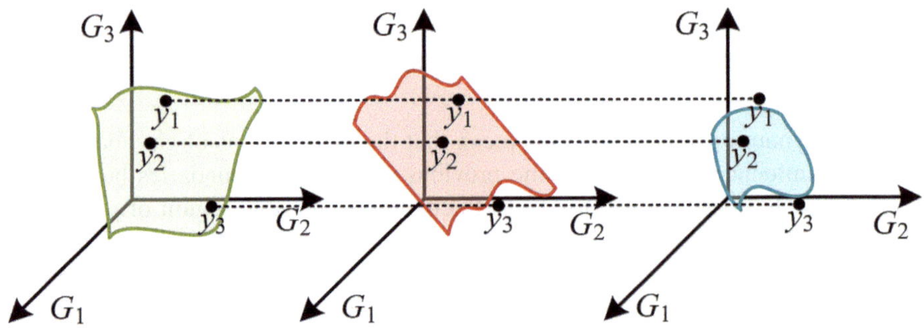

Fig. 4.6 Time-varying characteristics under heterogeneous energy flows, reprinted from [10], with permission from IEEE

- The decision commands at the current moment not only determine the unit outputs at the current moment but also affect the operational security at future moments.
- From the electric side, its permissible region at the coupling points (with GSs and HSs) continuously changes due to the energy flow dynamics.

To characterize such permissible regions of GSs and HSs, let's first make the following assumptions:

- The permissible region is assumed to be a continuous and closed space, meaning that binary variables introduced by devices such as line circuit breakers and energy storage are not considered in the modelling process.
- According to the simulation analysis (which will be shortly introduced), the timescale of the GTs is much shorter than that of the CHP units, GSs, and HSs. During adjacent scheduling intervals, the operational state of GTs has already reached a steady state. Therefore, only the static model of GTs is considered in the modelling process.

Given the above assumptions, the variables in GSs and HSs can be categorized into two types: controllable variables and state variables [10]. Controllable variables can be adjusted in practical applications, including the pressure and mass flow rate of gas sources in the GS, supply temperature of thermal sources in the HS, input fuel flow rate of the CHP units, and input gas flow rate and pressure of GTs. State variables are determined by the given control variables through system characteristics, including gas load pressure in the GS, node temperature and return temperature of heat loads in the HS, inlet and outlet temperatures, power supply, and boiler/steam turbine pressure of the CHP units; and output power of the GT. In summary, the controllable variables $\mathbf{X}_t$ and state variables $\mathbf{Y}_t$ at t can be stacked as

$$\begin{cases} \mathbf{X}_t = \left\{\mathbf{X}_t^{\mathrm{g}}, \mathbf{X}_t^{\mathrm{h}}\right\} \\ \mathbf{X}_t^{\mathrm{g}} = \left\{\mathbf{p}_{\mathrm{nd,sr},t}, \mathbf{q}_{\mathrm{nd,sr},t}, \mathbf{q}_{\mathrm{g},t}\right\} \\ \mathbf{X}_t^{\mathrm{h}} = \left\{\mathbf{T}_{\mathrm{nd,\,sr},t}^{\mathrm{s}}, \mathbf{m}_{\mathrm{B},t}\right\} \end{cases}, \tag{4.58}$$

$$\begin{cases} \mathbf{Y}_t = \left\{\mathbf{Y}_t^{\mathrm{g}}, \mathbf{Y}_t^{\mathrm{h}}\right\} \\ \mathbf{Y}_t^{\mathrm{h}} = \left\{\mathbf{T}_{\mathrm{nd},t}^{\mathrm{r}}, \mathbf{T}_{\mathrm{nd,ld},t}^{\mathrm{s}}, \mathbf{p}_{\mathrm{B},t}, \mathbf{p}_{\mathrm{T},t}, \mathbf{P}_{\mathrm{T},t}, \mathbf{T}_{\mathrm{SHE},t}^{\mathrm{o}}, \mathbf{T}_{\mathrm{SHE},t}^{\mathrm{i}}\right\} \\ Y_t^{\mathrm{g}} = \left\{\mathbf{p}_{\mathrm{nd,ld},t}, \mathbf{q}_{\mathrm{nd,ld},t}, \mathbf{P}_{\mathrm{g},t}\right\} \end{cases}, \tag{4.59}$$

where $\mathbf{X}^{\mathrm{h}}$ and $\mathbf{Y}^{\mathrm{h}}$ denote the controllable and state variables in the HS, respectively; $\mathbf{X}^{\mathrm{g}}$ and $\mathbf{Y}^{\mathrm{g}}$ denote the controllable and state variables in the GS, respectively.

Given the symbols above, a feasible region of a combination of one GS and HS at arbitrary time step t_j can be expressed as

$$\boldsymbol{\Omega} \triangleq \left\{\mathbf{X}_j \left| \begin{matrix} \mathbf{y}_{\mathrm{eq}}(\mathbf{X}_1,\cdots,\mathbf{X}_j,\mathbf{Y}_1,\cdots,\mathbf{Y}_j) \le \mathbf{0} \\ \mathbf{y}_{\mathrm{ieq}}(\mathbf{X}_1,\cdots,\mathbf{X}_j,\mathbf{Y}_1,\cdots,\mathbf{Y}_j) = \mathbf{0} \end{matrix} \right.\right\} \tag{4.60}$$

where $\mathbf{y}_{\mathrm{eq}}$ and $\mathbf{y}_{\mathrm{ieq}}$ are the sets of equality and inequality constraints in the IES. In the HS, $\mathbf{y}_{\mathrm{eq}}$ stack (4.33)–(4.36) and (4.40)–(4.56) while $\mathbf{y}_{\mathrm{ieq}}$ stack (4.37)–(4.38) and (4.48)–(4.53); In the GS, $\mathbf{y}_{\mathrm{eq}}$ stack (4.8), (4.11)–(4.12) while $\mathbf{y}_{\mathrm{ieq}}$ stack (4.13)–(4.14), (4.26) and the following (4.61):

$$q_{\mathrm{gt,min}} \le q_{\mathrm{gt}} \le q_{\mathrm{gt,max}}. \tag{4.61}$$

Remark 4.1 One GS and HS are coupled with transmission and distribution sub-systems through heterogeneous energy flows, which are included as the coupling variables in $\mathbf{X}_t$. Consequently, when all the controllable variables in $\mathbf{X}_t$ except the coupling variables are fixed, the feasible region in (4.60) becomes the exact permissible region of the combination of one GS and HS. ■

Remark 4.2 A transmission and distribution sub-system may be only interconnected with either GSs or HSs. In this context, only either the permissible region of one GS or the permissible region of one HS is required. Particularly, combining permissible regions of one GS and HS is equivalent to the permissible region of the combination of one GS and HS. This drives this chapter to further focus on resource aggregation for either one GS or HS in the coming Sect. 4.4. ■

4.4 Resource Aggregation for Heterogeneous Energy Flows

As demonstrated in the above contents, the permissible regions of one GS or one HS are dynamic nonlinear geometry, which results in intractable characterizations. To resolve this issue, Sect. 4.4 will exhibit an explicit version to get the permissible regions when employing certain simplifications whose resultant aggregation accuracy can still be desirable.

4.4.1 Permissible Region of Gas Systems

Characterizing the permissible region of one GS starts from handling a set of partial-differential algebraic equations, as delineated below.

According to [15, 16], (4.8) can be converted into the following algebraic equations:

$$\begin{aligned} p_{\mathrm{L},j} = &\sum_{n=1}^{\infty} e^{-g_n t_j} u_{n,0}(-1)^{n-1} - \frac{\lambda_g w L}{2DS} q_{\mathrm{L},j} + p_{0,j} \\ &+ \sum_{n=1}^{\infty} \frac{\lambda w E_{1n}}{2DS}(-1)^{n-1}\left(q_{\mathrm{L},j} - \frac{q_{\mathrm{L},0}}{e^{g_n t_j}} - \sum_{i=1}^{j} q_{\mathrm{L},i} \frac{e^{g_n \Delta t} - 1}{e^{g_n (j-i+1)\Delta t}} \right), \end{aligned} \tag{4.62}$$

$$\begin{aligned} q_{0,j} = &q_{\mathrm{L},j} + \frac{DS\pi}{\lambda_g w L} \sum_{n-1}^{\infty} (1-2n) u_{n,0} e^{-g_n t_j} \\ &+ \frac{\pi}{2L} \sum_{n-1}^{\infty} (1-2n) E_{1n} \left(q_{\mathrm{L},j} - \frac{q_{\mathrm{L},0}}{e^{g_{n,j}}} - \sum_{i=1}^{j} q_{\mathrm{L},i} \frac{e^{g_n \Delta t} - 1}{e^{g_n (j-i+1)\Delta t}} \right), \end{aligned} \tag{4.63}$$

where $p_{\mathrm{L},j}$ is the outlet pipeline pressure at time j; $q_{0,j}$ is the inlet pipeline mass flow rate at time j; $p_{0,j}$ is the inlet pipeline pressure at time j; $q_{\mathrm{L},j}$ is the outlet pipeline mass flow rate at time j; g_n, $u_{n,0}$ and E_{1n} are constant coefficients defined for simplification, whose expressions can be given as

$$\alpha = \frac{2Dc^2}{\lambda_g w}, \tag{4.64}$$

$$g_n = \frac{(2n-1)^2 \pi^2 \alpha}{4L^2}, \tag{4.65}$$

$$E_{1n} = \frac{8L(-1)^{n-1}}{(2n-1)^2 \pi^2}, \tag{4.66}$$

$$u_{n,0} = E_{1n}\left(K_1 + \frac{\lambda_g w}{2DS} q_{L,0}\right), \tag{4.67}$$

where K_1 is the coefficient of the initial condition of the pipeline pressure, $q_{L,0}$ is the initial condition of the pipeline mass flow rate.

By extending (4.62) and (4.63) from $t = 0$ to $t = j$ and re-organize them into a matrix form, we can obtain

$$\begin{bmatrix} \mathbf{p}_L \\ \mathbf{q}_0 \end{bmatrix} = \begin{bmatrix} \mathbf{H}_1 & \mathbf{H}_2 \\ \mathbf{H}_3 & \mathbf{H}_4 \end{bmatrix} \begin{bmatrix} \mathbf{p}_0 \\ \mathbf{q}_L \end{bmatrix} + \begin{bmatrix} \boldsymbol{\gamma}_p \\ \boldsymbol{\gamma}_q \end{bmatrix}, \tag{4.68}$$

$$\begin{cases} \mathbf{p}_0 = \left[p_{0,1}, \ldots, p_{0,N_t}\right] \\ \mathbf{p}_L = \left[p_{L,1}, \ldots, p_{L,N_t}\right] \\ \mathbf{q}_L = \left[q_{L,1}, \ldots, q_{L,N_t}\right] \\ \mathbf{q}_0 = \left[q_{0,1}, \ldots, q_{0,N_t}\right] \end{cases}, \tag{4.69}$$

where $\mathbf{p}_L$ and $\mathbf{q}_0$ are the vectors of the pipeline outlet pressure and inlet mass flow rate, respectively; $\mathbf{q}_L$ and $\mathbf{p}_0$ are the vectors of the pipeline inlet pressure and outlet mass flow rate, respectively; $\mathbf{H}_1$, $\mathbf{H}_2$, $\mathbf{H}_3$, and $\mathbf{H}_4$ are the transfer matrixes of gas states; $\boldsymbol{\gamma}_p$ and $\boldsymbol{\gamma}_q$ are the pressure and mass flow rate components determined by initial conditions.

Particularly, the elements in $\mathbf{H}_1$, $\mathbf{H}_2$, $\mathbf{H}_3$, $\mathbf{H}_4$, $\boldsymbol{\gamma}_p$ and $\boldsymbol{\gamma}_q$ are expressed as:

$$\gamma_{p,j} = \sum_{n=1}^{\infty} \left(u_{n,0} + E_{1n} m_0\right) (-1)^{n-1} e^{-g_n t_j}, \tag{4.70}$$

$$\gamma_{q,j} = \frac{DS\pi}{\lambda w L} \sum_{n-1}^{\infty} (1 - 2n) \left(u_{n,0} + E_{1n} m_0\right) e^{-g_n t_j}, \tag{4.71}$$

$$\mathbf{H}_1 = \mathbf{1}, \mathbf{H}_3 = \mathbf{0}, \tag{4.72}$$

$$\underset{(i=j)}{\mathbf{H}_{2,ji}} = \frac{\lambda_g w}{2DS} \left(\sum_{n=1}^{\infty} E_{1n} (-1)^{n-1} e^{-g_n \Delta t} - L\right), \tag{4.73}$$

$$\underset{(i<j)}{\mathbf{H}_{2,ji}} = \frac{\lambda_g w}{2DS} \left(\sum_{n=1}^{\infty} E_{1n} (-1)^{n-1} \frac{1 - e^{g_n \Delta t}}{e^{g_n (j-i+1)\Delta t}}\right), \tag{4.74}$$

$$\underset{(i=j)}{\mathbf{H}_{4,j}} = \frac{\pi}{2L} \sum_{n=1}^{\infty} E_{1n} (1 - 2n) e^{-g_n \Delta t} + 1, \tag{4.75}$$

$$\underset{(i<j)}{\mathbf{H}_{4,j}} = \frac{\pi}{2L}\sum_{n=1}^{\infty} E_{1n}(1-2n)\frac{1-e^{g_n \Lambda A}}{e^{g_n(j-i+1)\Delta t}}, \tag{4.76}$$

$$\underset{2,ji}{\mathbf{H}_{i>j}} = \underset{i,jji}{\mathbf{H}_{i>}} = \mathbf{0}. \tag{4.77}$$

To characterize the topological connections within the GS, the following incidence matrix is introduced:

- $\mathbf{A}_{\text{in}}$: $\mathbf{A}_{\text{in}}$ is used to relate the node mass flow rate with the pipeline outlet mass flow rate. If the outlet mass flow rate of pipeline j injects into node i, $a_{\text{in},ij} = 1$. Otherwise, $a_{\text{in},ij} = 0$.
- $\mathbf{A}_{\text{out}}$: $\mathbf{A}_{\text{out}}$ is used to relate the node mass flow rate with pipeline inlet mass flow rate. If the mass flow rate leaves from node i and then injects into pipeline j, $a_{\text{out},ij} = 1$. Otherwise, $a_{\text{out},ij} = 0$.
- $\mathbf{A}_{\text{nb1}}$: $\mathbf{A}_{\text{nb1}}$ is used to relate the node pressure with pipeline outlet pressure. If the pressure of node i equals the outlet pressure of pipeline j, $a_{\text{nb1},ij} = 1$. Otherwise, $a_{\text{nb1},ij} = 0$.
- $\mathbf{A}_{\text{nb2}}$: $\mathbf{A}_{\text{nb2}}$ is used to relate the node pressure with pipeline inlet pressure. If the pressure of node i is equal to the inlet pressure of pipeline j, $a_{\text{nb2},ij} = 1$. Otherwise, $a_{\text{nb2},ij} = 0$

According to the topological connections, (4.11) and (4.12) can be rewritten in the following matrix form:

$$\mathbf{A}_{\text{in}}\boldsymbol{q}_{\text{L}} - \mathbf{A}_{\text{out}}\boldsymbol{q}_{0} = \boldsymbol{q}_{\text{nd}}, \tag{4.78}$$

$$\boldsymbol{p}_{\text{L}} = \mathbf{A}_{\text{nb1}}\boldsymbol{p}_{\text{nd}}, \ \boldsymbol{p}_{0} = \mathbf{A}_{\text{nb2}}\boldsymbol{p}_{\text{nd}}. \tag{4.79}$$

Substituting (4.79) into the first row in (4.68) yields

$$\boldsymbol{q}_{\text{L}} = \mathbf{H}_2^{-1}\left((\mathbf{A}_{\text{nb1}} - \mathbf{A}_{\text{nb2}})\,\boldsymbol{p}_{\text{nd}} - \boldsymbol{\gamma}_{\text{p}}\right) = \mathbf{H}_{\text{a1}}\boldsymbol{p}_{\text{nd}} + \mathbf{H}_{\text{a2}}. \tag{4.80}$$

Substituting (4.79) and (4.80) into the second row in (4.68) yields

$$\begin{aligned} \boldsymbol{q}_{0} &= \mathbf{H}_4\left(\mathbf{H}_{\text{a1}}\boldsymbol{p}_{\text{nd}} + \mathbf{H}_{\text{a2}}\right) + \gamma_{\text{q}} \\ &= \mathbf{H}_4\mathbf{H}_{\text{a1}}\boldsymbol{p}_{\text{nd}} + \mathbf{H}_4\mathbf{H}_{\text{a2}} + \gamma_{\text{q}}. \\ &= \mathbf{H}_{\text{a3}}\boldsymbol{p}_{\text{nd}} + \mathbf{H}_{\text{a4}} \end{aligned} \tag{4.81}$$

Furthermore, substituting (4.80) and (4.81) into (4.78) yields

$$\begin{aligned} \boldsymbol{q}_{\text{nd}} &= (\mathbf{A}_{\text{in}}\mathbf{H}_{\text{a1}} - \mathbf{A}_{\text{out}}\mathbf{H}_{\text{a3}})\,\boldsymbol{p}_{\text{nd}} + (\mathbf{A}_{\text{in}}\mathbf{H}_{\text{a2}} - \mathbf{A}_{\text{out}}\mathbf{H}_{\text{a4}}) \\ &= \mathbf{H}_{\text{a5}}\boldsymbol{p}_{\text{nd}} + \mathbf{H}_{\text{a6}} \end{aligned}. \tag{4.82}$$

Note that (1) $\mathbf{H}_{a1}$, $\mathbf{H}_{a3}$, and $\mathbf{H}_{a5}$ are constant coefficient matrixes in the above equations, (2) $\mathbf{H}_{a2}$, $\mathbf{H}_{a4}$, and $\mathbf{H}_{a6}$ are constant coefficient vectors in the above equations. Exploring (4.82) in a matrix form, we can obtain the following equation:

$$\begin{bmatrix} \boldsymbol{q}_{\text{nd,sr}} \\ \boldsymbol{q}_{\text{nd,ns}} \end{bmatrix} = \begin{bmatrix} \mathbf{H}_{a51} & \mathbf{H}_{a52} \\ \mathbf{H}_{a53} & \mathbf{H}_{a54} \end{bmatrix} \begin{bmatrix} \boldsymbol{p}_{\text{nd,sr}} \\ \boldsymbol{p}_{\text{nd,ns}} \end{bmatrix} + \begin{bmatrix} \mathbf{H}_{a61} \\ \mathbf{H}_{a62} \end{bmatrix}, \tag{4.83}$$

where $\mathbf{H}_{a51}$, $\mathbf{H}_{a52}$, $\mathbf{H}_{a53}$, and $\mathbf{H}_{a54}$ are constant coefficient matrixes; $\mathbf{H}_{a61}$ and$\mathbf{H}_{a62}$ are constant coefficient vectors; subscript ns is the symbol of variables at non-source nodes.

From (4.83), we can obtain the following explicit equations:

$$\boldsymbol{p}_{\text{nd,ss}} = \mathbf{H}_{as4}^{-1}\left(\boldsymbol{q}_{\text{nd,ns}} - \mathbf{H}_{as3}\boldsymbol{p}_{\text{nd,sr}} - \mathbf{H}_{a62}\right), \tag{4.84}$$

$$\boldsymbol{q}_{\text{nd, sy}} = \mathbf{H}_{a51}\boldsymbol{p}_{\text{nd, sr}} + \mathbf{H}_{as2}\boldsymbol{p}_{\text{nd, ns}} + \mathbf{H}_{a61}. \tag{4.85}$$

Finally, the GS part of the feasible region in (4.60) can be formulated as a polytope using the intersection of (4.13), (4.14), (4.26), (4.84) and (4.85). By further fixing the controllable variables except those which interact with the transmission/distribution sub-subsystems, the polytope above becomes the permissible region of one GS.

4.4.2 Permissible Region of Heating Systems

For the quality and quantity regulation modes, the former one has been widely adopted in North China, Russia, Norway, Netherlands, etc. Consequently, the permissible region relevant to HSs is discussed under the quality regulation mode. In this condition, the permissible region of HSs is governed by the high-dimensional PDAEs. According to [7, 17], (4.33) can be converted into the following algebraic equation:

$$\begin{aligned} T_{x,t} &= (\psi_{x-vt})\, K_t\left[\delta(t) - \delta(t-\xi_x)\right] \\ &+ \left(\varphi_{t-\xi_x}\right) J_x \delta(t-\xi_x) \end{aligned}, 0 \le x \le L_i, i \in \Phi_{\text{h}}, \tag{4.86}$$

$$K_{i,t} = e^{\frac{-v_i, t}{C_{\text{w}} m_i \lambda_{h,i}}} \quad i \in \Phi_{\text{h}}, \tag{4.87}$$

$$J_{i,x} = e^{\frac{-x}{C_{\text{w}} m_i \lambda_{g,i}}} \quad i \in \Phi_{\text{h}}, \tag{4.88}$$

$$\xi_{i,x} = \frac{x}{v_i} \quad i \in \Phi_{\text{h}}, \tag{4.89}$$

$$\psi_{x-vt} = \psi_{x-t}^{\text{abs}} - T_{\text{a},t}, \varphi_{t-\xi_t} = \varphi_{t-\xi_x}^{\text{abs}} - T_{\text{a},t}, \tag{4.90}$$

where $\delta(t)$ is the unit stepped function[3]; ψ_x is the relative initial condition regarding x, corresponding to the pipe temperature distribution at $t = 0$; φ_t is the relative boundary condition regarding t, corresponding to the pipe inlet temperature series at $x = 0$; ψ_x^{abs} is the absolute initial condition; ψ_t^{abs} is the absolute boundary condition; ξ_x denotes the transfer delay between the temperatures at a distance x and the pipeline outlet; K_t and J_x correspond to the transfer coefficients of the initial and boundary conditions, respectively; K_t reflects the transfer loss when the temperature at a distance x transfers to the pipeline outlet; J_x reflects the transfer loss when the temperature transfers from the pipeline inlet to the outlet.

Since the mass flow rate in one quality-regulated HS is fixed and the temperature is transferred through water flow, the temperature transfer paths between the source and load nodes in the HS are fixed. For an arbitrary load node, its temperature contains two components. The first component is determined by the temperature of the source node (i.e. boundary condition) after a certain delay, which is defined as T_φ, and the second component is determined by the initial temperature that finally flows into the node along pipelines, which is defined as T_Ψ. For simplification, the following definitions are introduced:

- Φ_{ij}^{h} is defined as the set of temperature transfer paths between nodes i and j, $\Phi_{ij,n}^{\mathrm{h}}$ is defined as the nth path in Φ_{ij}^{h}.
- ξ_{ij}^n is defined as the delay along the nth path between nodes i and j.
- D_{ij}^n is defined as the diversion ratio of temperature transfer along $\Phi_{ij,n}^{\mathrm{h}}$. D_{ij}^n is constant and expressed as

$$D_{ij}^n = \prod_b \frac{m_b}{\sum_{k\in\Phi_{\mathrm{d}}^{\mathrm{os}}} m_k + m_{\mathrm{nd},e}}, \; b \in \Phi_{ij}^{\mathrm{h}}, e \in \Theta_{ij,b}^{\mathrm{h}}, \tag{4.91}$$

where Θ_{ij}^{h} is the set of nodes along Φ_{ij}^{h}.

- R_{ij}^n is defined as the transfer coefficient along $\Phi_{ij,n}^{\mathrm{h}}$ and can be expressed as

$$R_{ij}^n = \prod J_{k,L_k} \quad i, j \in \Theta^{\mathrm{h}}, k \in \Phi_{ij,n}^{\mathrm{h}}. \tag{4.92}$$

With the above definitions, T_φ and T_Ψ at arbitrary load is expressed as

$$T_{\mathrm{nd},j,t}^{\varphi} = \sum_{i\in\Theta_{\mathrm{sr}}^{\mathrm{h}}} \sum_{k\in\Phi_{ij}^{\mathrm{h}}} D_{ij}^k R_{ij}^k \left(T_{\mathrm{nd},i,t-\xi_{ij}^k} - T_{\mathrm{a},t}\right), \; j \in \Theta_{\mathrm{h,\,ns}}, \; t - \xi_{ij}^k > 0, \tag{4.93}$$

$$\begin{gathered} T_{\mathrm{nd},j,t}^{\Psi} = \sum_{i\in\Theta_{\mathrm{sr}}^{\mathrm{h}}} \sum_{k\in\Phi_{\mathrm{ij}}^{\mathrm{h}}} D_{ij}^k \sum_{n\in\Phi_{i,j}^k} \left(\psi_{n,L_n - v_n\left(t-\xi_{n-j}^k\right)} - T_{\mathrm{a},t}\right) K_{n,v_n\left(t-\xi_{n-j}^k\right)} R_{n-j}^k, \\ j \in \Theta_{\mathrm{ld}}^{\mathrm{h}}, 0 \le v_n\left(t - \xi_{n-j}^k\right) \le L_n \end{gathered} \tag{4.94}$$

[3] If $t \ge 0$, $\delta(t) = 1$. Otherwise, $\delta(t) = 0$.

$$T_{\mathrm{nd},j,t} = T^{\psi}_{\mathrm{nd},j,t} + T^{\varphi}_{\mathrm{nd},j,t}, \quad T^{\mathrm{abs}}_{\mathrm{nd},j,t} = T_{\mathrm{nd},j,t} + T_{\mathrm{a},t}, \tag{4.95}$$

where n^- is the outlet node of pipeline n.

Finally, the HS part of the feasible region in (4.60) can be formulated as a polytope using the intersection of (4.36)–(4.38), (4.57), and (4.93)–(4.95). By further fixing the controllable variables except those which interact with the transmission/distribution sub-subsystems, the polytope above becomes the permissible region of one HS.

4.4.3 Verification Method

To verify the effectiveness of the permissible region, a TSCH method will be introduced as the reference, as detailed below.

- Step 1. Randomly generate sufficient sets of controllable variables $\mathbf{X}_j$ within their intervals $\left[\mathbf{X}_j^{\min}, \mathbf{X}_j^{\max}\right]$.
- Step 2. Select a set of generated $\mathbf{X}_j$ and then perform the dynamic simulation to obtain the state variables $\mathbf{Y}_j$.
- Step 3. Check whether $\left(\mathbf{X}_j, \mathbf{Y}_j\right)$ satisfies the constraints in Sect. 4.2. If so, let's add $\left(\mathbf{X}_j, \mathbf{Y}_j\right)$ into the set of secure points.
- Step 4. Repeat Steps 2 and 3 until all $\mathbf{X}_j$ are traversed and checked.
- Step 5. Compute the convex hull of secure points.

According to the steps above, the results from the TSCH method with sufficient simulations are recognized as the accurate result (denoted as $\mathbf{\Omega}_{\mathrm{ref}}$ hereafter), i.e., the points within $\mathbf{\Omega}_{\mathrm{ref}}$ are secure. Meanwhile, points out of $\mathbf{\Omega}_{\mathrm{ref}}$ are insecure. On this basis, the following two metrics can be introduced to evaluate the aggregation results from Sects. 4.4.1 and 4.4.2 (denoted as $\mathbf{\Omega}$):

- Accurate degree: This metric checks how many points in $\mathbf{\Omega}$ are truly secure (i.e., fulfilling the constraints in Sect. 4.2).
- Conservative degree: This metric checks the coverage of the methods in Sects. 4.4.1 and 4.4.2.

The detailed steps to compute the accurate and conservative degrees and their expressions are given below.

- Step 1. Divide the intervals $\left[\mathbf{X}_j^{\min}, \mathbf{X}_j^{\max}\right]$ of $\mathbf{X}_j$ equally into N points (N is sufficiently large, e.g., N = 1000M for a M-dimensional feasible region).
- Step 2. Count the number of points within $\mathbf{\Omega}_{\mathrm{ref}}$ and denote it as num_4.
- Step 3. Count the number of points within both $\mathbf{\Omega}_{\mathrm{ref}}$ and $\mathbf{\Omega}$, and denote it as num_1.
- Step 4. Count the number of points within $\mathbf{\Omega}$, and define it as num_2. The accurate degree equals $\mathrm{num}_1/\mathrm{num}_2$, which reflects the coverage proportion of $\mathbf{\Omega}_{\mathrm{ref}}$. For example, if the accurate degree equals D = 97, 97% of $\mathbf{\Omega}$ are located within $\mathbf{\Omega}_{\mathrm{ref}}$.

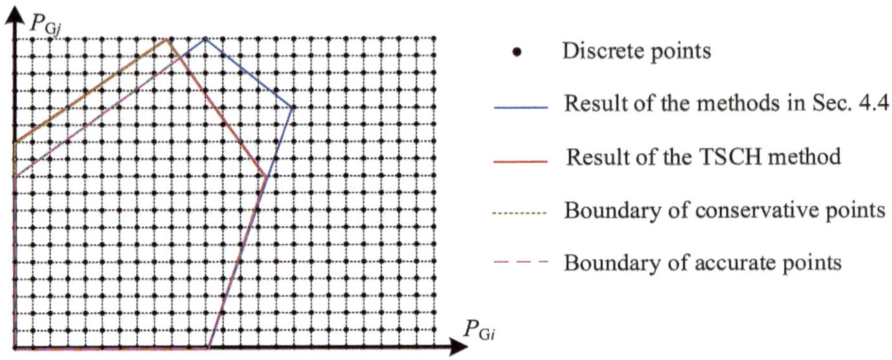

Fig. 4.7 Computational illustration of the accurate and conservative degrees, reprinted from [10], with permission from IEEE

- Step 5. Count the number of points within $\mathbf{\Omega}_{\text{ref}}$ but out of $\mathbf{\Omega}$, and denote it as num_3. The conservative degree equals $\text{num}_3/\text{num}_4$, which reflects the proportion of $\mathbf{\Omega}_{\text{ref}}$ that the proposed method cannot cover. For example, if the conservative degree equals 2, 2% of $\mathbf{\Omega}_{\text{ref}}$ are out of $\mathbf{\Omega}$.

A verification illustration is given in Fig. 4.7.

4.5 Case Studies

4.5.1 Multi-timescales of Heterogeneous Energy Flows

To illustrate the timescale differences in GSs and HSs, Sect. 4.5.1 introduces several simulations based on several well-known cases.

For the HS dynamics, the Luhua HS is used as an illustrative example, which is a real-world system in China. Its parameters can be found in [20]. The spatial and temporal steps are set to be 92.5 m and 15 min, respectively. The ambient temperature is 0^{circ}C, and the simulation period is 24 h. The simulation results are shown in Fig. 4.8. It can be found that a delay between the inlet and outlet temperatures is up to around 92 min. This shows the timescale of tens of minutes in the HS that cannot be neglected when performing its resource aggregation.

For the GS dynamics, a single gas pipeline from GasLib is utilized as an illustrative example [21]. Its parameters can be found in [22]. The spatial and temporal steps are set to be 500 m and 6 min, respectively. The simulation period is 6 h. The simulation results are shown in Fig. 4.9. It can be found that the delays between the inlet and outlet pressures, and inlet and outlet mass flow rates are around 60 min. This shows the timescale in the GS is at the level of tens of minutes, which cannot be neglected when performing its resource aggregation.

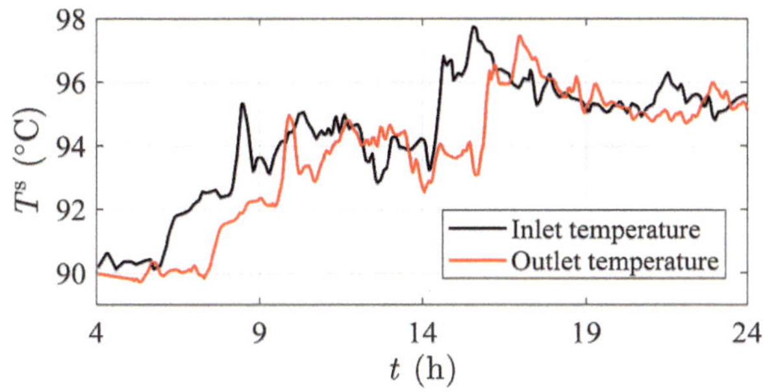

Fig. 4.8 Illustration of slow dynamics in the HS, reprinted from [16], open access

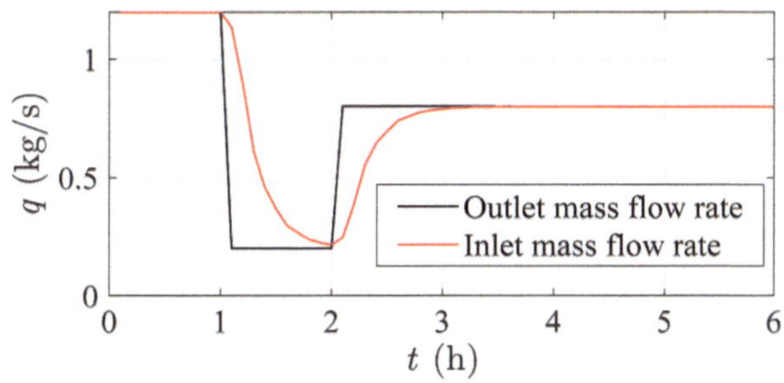

(a) Mass flow rate.

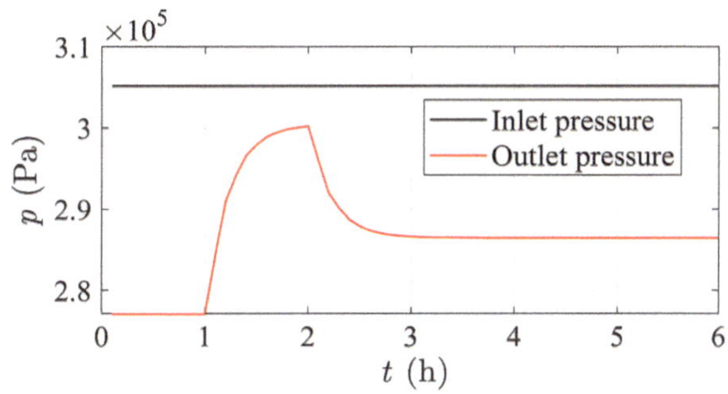

(b) Pressure.

Fig. 4.9 Illustration of slow dynamics in the GS, reprinted from [16], open access

4.5.2 Accuracy Verification

Section 4.5.2 will employ two cases to investigate the accuracy of the feasible region of one combination of HSs and GSs.[4] Case I is composed of a 9-bus transmission sub-system and a 2-node HS, which are coupled through a CHP unit at Bus 3 in the transmission sub-system and Node 1 in the HS, as shown in Fig. 4.10a. Case II is a modified HE-IES containing the modified 30-bus transmission sub-system and the 60-node HS, as shown in Fig. 4.10b. The HS is a real-world system in Tianjin,

[4] Since the permissible region of one combination of HSs and GSs is the feasible region of one combination of HSs and GSs under given controllable variables, the accuracy of the permissible region is examined through the feasible region in (4.60).

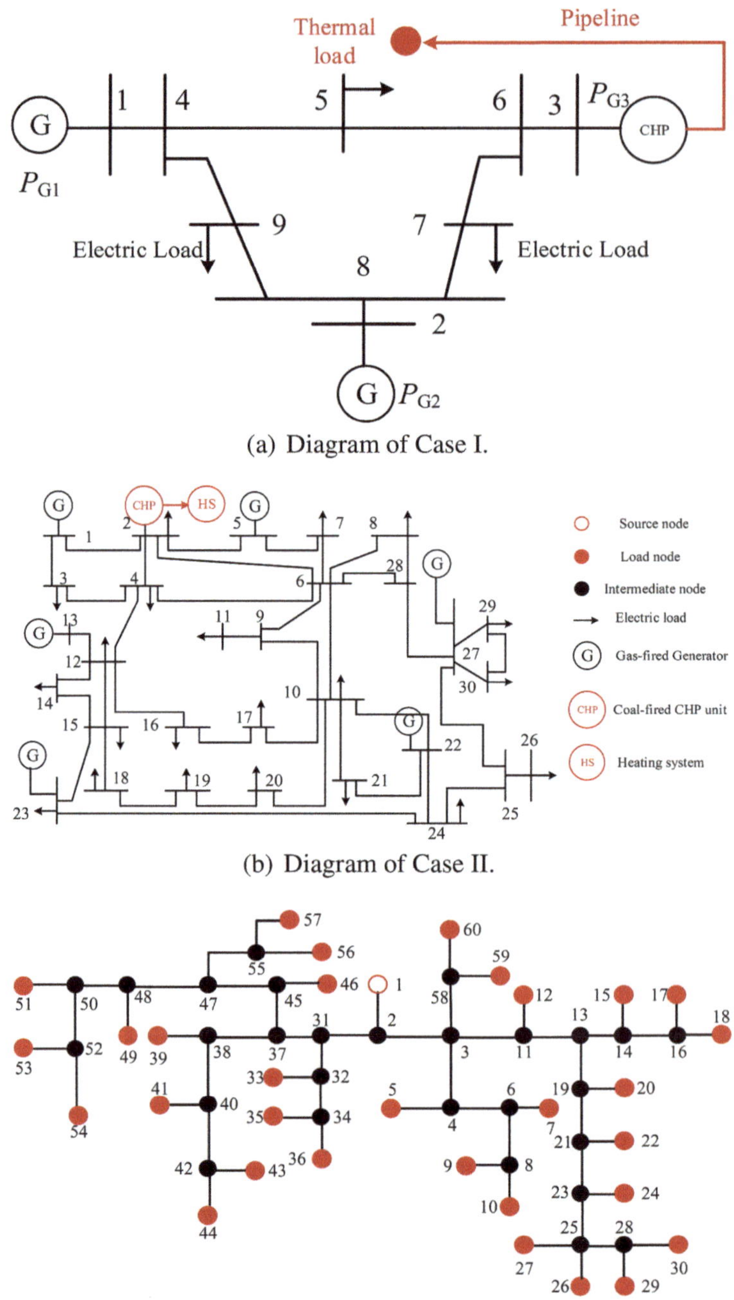

(a) Diagram of Case I.

(b) Diagram of Case II.

(c) HS diagram in Case II.

Fig. 4.10 Diagrams of test systems, reprinted from [10], with permission from IEEE

China, as shown in Fig. 4.10c. In Case II, the transmission sub-system and the HS are coupled through a CHP unit at Bus 2 in the transmission sub-system and Node 1 in the HS. All the parameters and data are available in [10].

In Case I, the maximum time delays for the CHP unit and HS are 31 and 93 min, respectively. As for Case II, the maximum time delays for the CHP unit and HS are 18 and 91 min, respectively. The time step sizes in Cases I and II are both five minutes. To implement the TSCH method for verification, 3×10^5 random simulations are conducted at each time step in Case I, and 7×10^5 simulations at each time step in Case II to ensure the reference accuracy. The accuracy verification is then conducted using 1×10^6 operational points to compute the accurate and conservative degrees.

The feasible regions of Cases I and II are visualized in Figs. 4.11 and 4.12, with the areas and computational times summarized in Tables 4.1 and 4.2. The performance metrics are provided in Table 4.3. Despite minor mismatches in some parts of the boundaries, the results of the methods in Sect. 4.4 show a desirable accuracy. The value of the accurate degree of the methods in Sect. 4.4 for Cases I and II are approximately 96.0 and 97%, respectively. This indicates that their major parts are feasible to the constraints in Sect. 4.2. The mismatches are primarily due to the error margin with certain simplification in Sect. 4.4, which can introduce characterization errors of the feasible region. These factors lead to differences in the boundary characterization compared to that of the TSCH method, as seen in Fig. 4.11a and d. The conservative degree in Cases I and II equal around 5 and 2%, respectively.

Computationally, the methods in Sect. 4.4 offer significant advantages compared with the TSCH method, being approximately 100 times more efficient. The TSCH method requires extensive simulations to build the feasible region at each time step, involving repeated solutions to the nonlinear constraints in Sect. 4.2, which is time-consuming. In contrast, the methods in Sect. 4.4 only compute a certain of hyperplane coefficients, eliminating the need for repeated simulations and optimizations. This makes the methods in Sect. 4.4 highly efficient for real-time applications.

4.5.3 Influence of Dynamics in Energy Flows

Section 4.5.3 will take one HS and one transmission sub-system as examples to explore the impacts of integrating heterogeneous energy flows on power systems under the following three scenarios:

S1: Only the power flow constraints are considered and capacity constraints of CHP units

S2: All the constraints in power systems and CHP units are considered.

S3: All the constraints in Sect. 4.2 are considered.

Particularly, a 5 min time step is set over a 6 h horizon. The simulation results are provided in Fig. 4.13 and summarized in Table 4.4. According to the simulations, it can be found that the integration of CHP units and one HS affects power systems

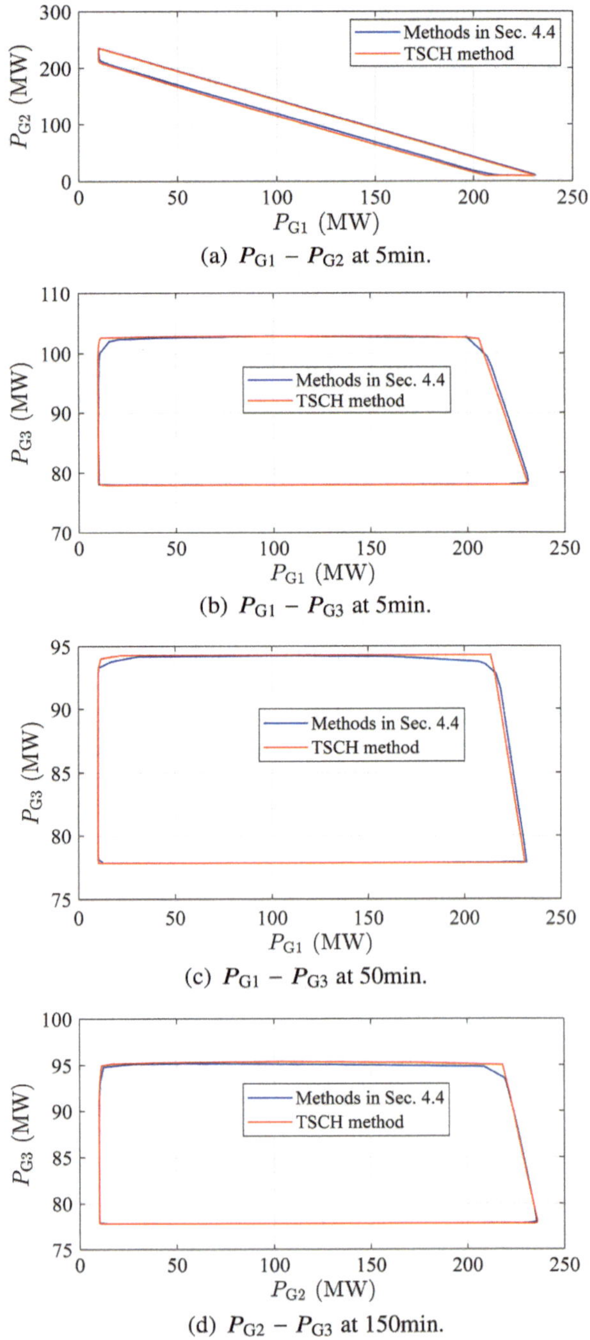

Fig. 4.11 Comparison between the feasible regions of the TSCH method and the methods in Sect. 4.4 for Case I, reprinted from [10], with permission from IEEE

Fig. 4.12 Comparison between the feasible regions of the TSCH method and the methods in Sect. 4.4 for Case II, reprinted from [10], with permission from IEEE

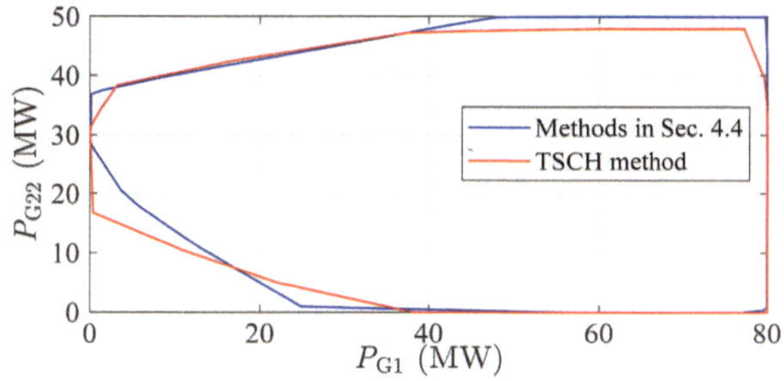

(a) $P_{G1} - P_{G22}$ at 55min.

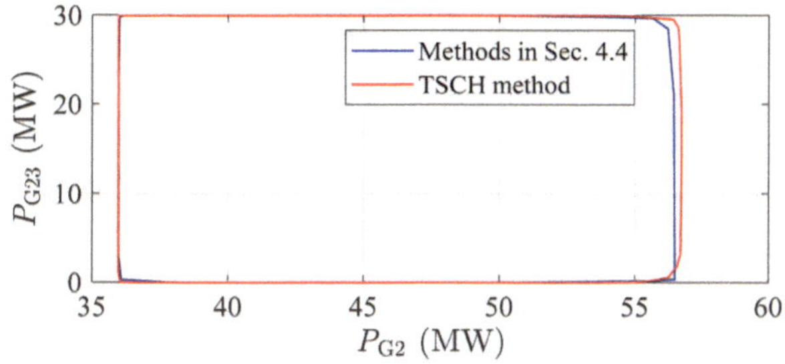

(b) $P_{G2} - P_{G23}$ at 5min.

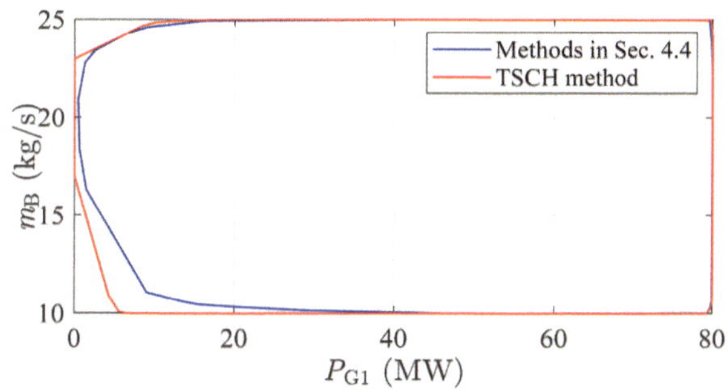

(c) $P_{G1} - m_B$ at 50min.

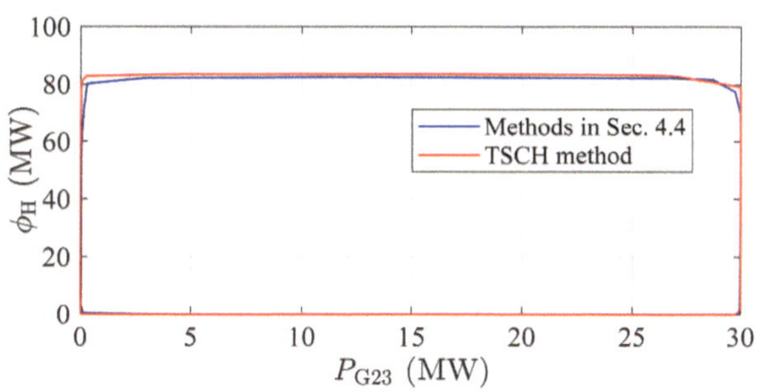

(d) $P_{G23} - \phi_H$ at 150min.

Table 4.1 2-D feasible regions of the TSCH method and the methods in Sect. 4.4 for Cases I and II, reprinted from [10], with permission from IEEE

Case		Feasible region of Sect. 4.4 $(MW)^2$			Feasible region of TSCH $(MW)^2$		
		5 min	50 min	150 min	5 min	50 min	150 min
I	$P_{G1} - P_{G2}$	5103	3444	3634	5246	3595	3794
	$P_{G1} - P_{G3}$	5144	3473	3663	5157	3496	3699
	$P_{G2} - P_{G3}$	5230	3531	3724	5271	3575	3783
II	$P_{G1} - P_{G22}$	3374	3438	3407	3328	3314	3296
	$P_{G2} - P_{G22}$	1009	1008	1072	984	976	1026
	$P_{G2} - P_{G23}$	612	610	646	621	616	649

Table 4.2 Computational time of the TSCH method and the methods in Sect. 4.4 for Cases I and II, reprinted from [10], with permission from IEEE

Case	Horizon (min)	Methods in Sect. 4.4 (s)	TSCH method (s)
I	5	3.4	706.6
	50	4.5	738.7
	150	5.7	723.2
II	5	7.3	24321
	50	6.7	23780
	150	7.2	23793

Table 4.3 Performance metrics of the methods in Sect. 4.4 for Cases I and II, reprinted from [10], with permission from IEEE

Degree	Accurate degree			Conservative degree		
I	5 min	50 min	150 min	5 min	50 min	150 min
	95.6%	96.8%	96.0 %	4.91%	5.07%	5.04%
II	5 min	50 min	150 min	5 min	50 min	150 min
	97.8%	97.0%	96.3%	2.18%	1.96%	2.21%

in two primary ways observed from the power injection region[5]: the time-varying characteristics and the reshaping.

Initially, the power injection region of the transmission sub-system is static under given operational conditions. However, its power injection regions in scenarios S2 and S3 become time-dependent. In scenario S2, the variability is driven by the dynamics within the CHP units. As clarified above, the current state of the CHP units is influenced by its previous states, necessitating continuous adjustments in fuel flow

[5] The power injection region of one transmission sub-systems delineates the power injection set feasible to the constraints in the transmission sub-system.

Fig. 4.13 Power injection regions in scenarios S1–S3, reprinted from [10], with permission from IEEE

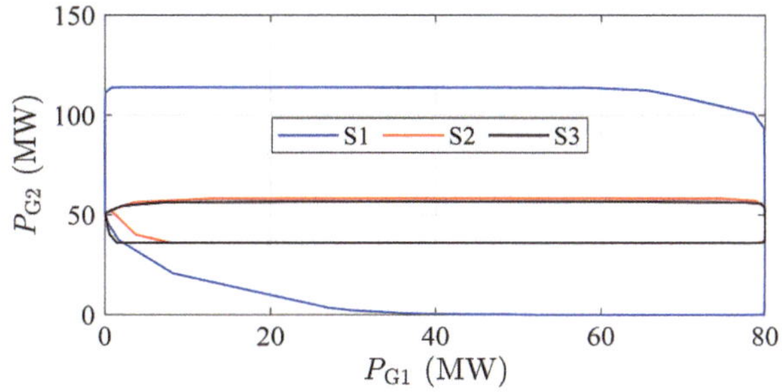

(a) $P_{G1} - P_{G2}$ at 50min.

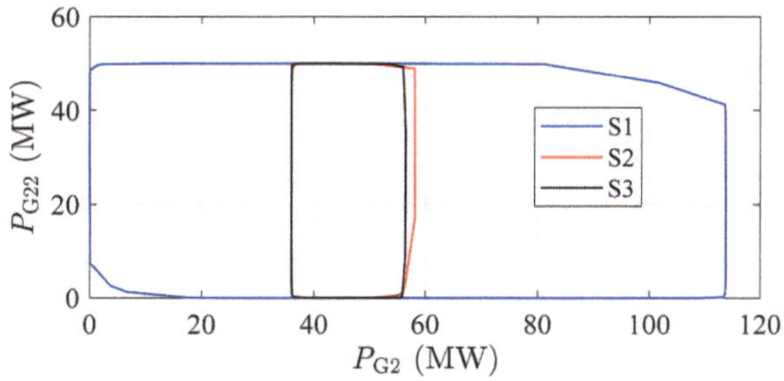

(b) $P_{G2} - P_{G22}$ at 50min.

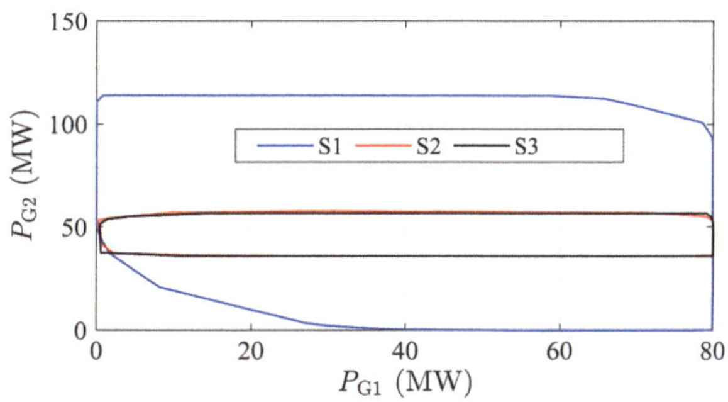

(c) $P_{G1} - P_{G2}$ at 150min.

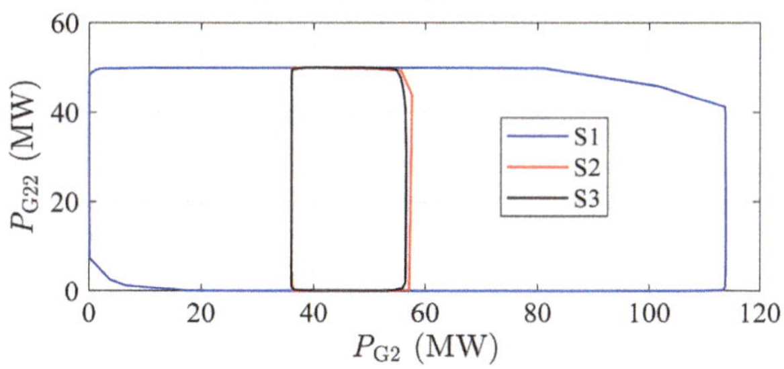

(d) $P_{G2} - P_{G23}$ at 2h.

Table 4.4 Volumes of power injections regions in scenarios S1–S3, reprinted from [10], with permission from IEEE

Domain	Horizon	S1/(MW)^2	S2/(MW)^2	S3/(MW)^2
$P_{G1} - P_{G2}$	5 min	8468	1694	1588
	50 min	8468	1674	1588
	150 min	8468	1674	1711
	3 h	8468	1674	1711
$P_{G22} - P_{G23}$	5 min	1499	1495	1494
	50 min	1499	1493	1498
	150 min	1499	1492	1490
	3 h	1499	1492	1490

and pressure to maintain outputs during the early stages. Therefore, the power injection region at 5 min differs from that at 50 min in scenario S2. As the simulation progresses and reaches the maximum delay (18 min), the states in the CHP units stabilize, leading to a consistent power injection region. Consequently, the power injection regions at 50, 150 min, and 3 h in scenario S2 are almost identical.

When the transmission sub-system is further integrated with one HS in scenario S3, the thermal and electrical outputs of the CHP units are constrained by the HS, adding complexity to the power injection region. After the CHP units reach their maximum delays, the HS dynamics continue to influence the power injection region, causing differences at 50 and 150 min in scenario S3. Considering the maximum delay in this case is 109 min, the power injection region at 150 min reaches a critical state and stabilizes.

Regarding the geometric shape, the integration of CHP units and HS also reshapes the power injection region of the transmission sub-system according to their corresponding connections. The CHP units must balance pressure and temperature, requiring a certain level of thermal outputs, which reduces the adjustable power outputs at P_{G2}. Therefore, the power injection region in the domain of P_{G1} and P_{G2} in scenario S2 is smaller than that in scenario S1, as shown in Fig. 4.13a and c. As time goes on, the CHP unit continuously adjusts the input fuel flow and pressure to maintain output power. The more fuel flow at the previous moment, the larger the adjustment capability for the current electrical power output and the corresponding power injection region, and vice versa.

In the early horizon of scenario S3 defined as Period 1, the dynamics of the CHP unit are the primary influence on the power injection, similar to those in scenario S2. However, the CHP unit in scenario S3 must also meet specific thermal power and temperature requirements in the HS. This results in a smaller adjustable power output and power injection region in scenario S3 compared to those in scenario S2, as indicated in Table 4.4. After the maximum delay in the CHP unit reaches (18 min), the HS dynamics become the dominant factor affecting the time-varying power injection region. In this period defined as Period 2, the HS states are influenced by both the initial conditions and the CHP unit's thermal output. Initially, the impact of the HS's starting temperature diminishes, and the CHP unit must increase thermal output to meet the HS0's requirements, reducing the adjustable P_{G2}. Later in Period 2, the CHP units can reduce thermal power outputs due to the higher initial conditions (historical temperature distribution in the HS). Consequently, the power injection region at 150 min is larger than those at 5 and 50 min. The integration of one HS provides additional adjustments for the secure operations of the transmission sub-system during a specific period and delays the time-varying process. When the HS reaches its maximum delay, the initial conditions no longer affect the operations, and the power injection region stabilizes. This period is defined as Period 3, as shown by the comparison between the power injection regions at 150 min and 3 h.

In conclusion, the integration of CHP units significantly reduces the areas of the power injection region of the transmission sub-system. However, the integration of one HS provides an additional security margin for the transmission sub-system during a specific period and delays the time-varying process. Simulations suggest that a critical state of the power injection region exists after the longest dynamic process, i.e., the maximum time delay. This critical state reflects the long-term impact of CHP unit and HS integration and remains stable over time.

4.6 Summary

This chapter studies the resource aggregation for GSs and HSs which considers the dynamics of their energy flows. The core of this lies in incorporating the dynamic characteristics of multi-energy flows and coupling components into the modeling of the permissible region for heterogeneous energy flows (i.e., gas flows in GSs and heating flows in HSs) at the coupling points. Then, to ensure their computational efficiency, the closed-form expressions of the permissible regions are formulated under certain simplifications whose resultant aggregation accuracy can be still desirable. According to theoretical analysis and numerical experiments in this chapter, the following conclusions can be drawn:

- Multi-energy coupling at the equipment level is jointly manifested by state quantities such as power, pressure, and temperature. Detailed operational characteristic models of equipment add additional constraints regarding state quantities - such as pressure and flow at coupling nodes - and significantly shrink the power injection region of transmission and distribution sub-systems.
- Although the coupling of multi-energy networks continuously reduces the power injection region from the electric side, it can provide additional potential adjustments for renewable accommodation in low-carbon power systems during specific periods.

- The interactions with gas flows and heating flows endow transmission and distribution sub-systems with a time-varying process. Particularly, such a process can be divided into two stages where the first stage comes from equipment dynamics and the second stage arises from network dynamics. Their durations are consistent with the maximum delay at the equipment and the maximum delay of flow transmission, respectively.

References

1. Ramsebner, J., Haas, R., Auer, H., et al.: From single to multi-energy and hybrid grids: historic growth and future vision. Renew. Sustain. Energy Rev. **151**, 111520 (2021)
2. Wu, J., Yan, J., Jia, H., et al.: Integrated energy systems. Appl. Energy **167**, 155–157 (2016)
3. Zhang, S., Gu, W., Zhang, X., et al.: Dynamic modeling and simulation of integrated electricity and gas systems. IEEE Trans. Smart Grid **14**(2), 1011–1026 (2023)
4. Moeini, M., Abbaspour, A., Fotuhi, M., Hajipour, E.: A decomposed solution to multiple-energy carriers optimal power flow. IEEE Trans. Power Syst. **29**(2), 707–716 (2014)
5. Liu, X., Wu, J., Jenkins, N., Bagdanavicius, A.: Combined analysis of electricity and heat networks. Appl. Energy **162**, 1238–1250 (2016)
6. Gu, W., Wang, J., Lu, S., et al.: Optimal operation for integrated energy system considering thermal inertia of district heating network and buildings. Appl. Energy **199**, 234–246 (2017)
7. Zhang, S., Gu, W., Zhang, X., et al.: Fully analytical model of heating networks for integrated energy systems. Appl. Energy **327**, 120081 (2022)
8. Wang, W., Liu, J., Zeng, D., Fang, F., Niu, Y.: Modeling and flexible load control of combined heat and power units. Appl. Therm. Eng. **166**, 114624 (2020)
9. Zhang, S., Gu, W., Zhang, X., et al.: Towards fast and robust simulation in integrated electricity and gas system: a sequential united method. IEEE Trans. Power Syst. **39**(1), 1822–1836 (2024)
10. Zhang, S., Gu, W., Wang, J., et al.: Steady-state security region of integrated energy system considering thermal dynamics. IEEE Trans. Power Syst. **39**(2), 4138–4153 (2024)
11. Zhang, S., Pan G., Li B., et al.: Multi-timescale security evaluation and regulation of integrated electricity and heating system. IEEE Trans. Smart Grid, Early Access (2024)
12. Ailer, P., Santa, I., Szederkenyi, G., Hangos, K.: Nonlinear model-building of a low-power gas turbine. Period. Polytech. Transp. Eng. **19**(12), 117–135 (2001)
13. Sadeghian, O., Mohammadpour, A., Mohammadi, B., Sadiq, R., Hewage, K.: Risk-averse maintenance scheduling of generation units in combined heat and power systems with demand response. Reliab. Eng. Syst. Saf. **216**, 107960 (2021)
14. Chen, S., Wei, Z., Sun, G., Wei, W., Wang, D.: Convex hull based robust security region for electricity-gas integrated energy systems. IEEE Trans. Power Syst. **34**(3), 1740–1748 (2019)
15. Zhang, S., Chen, S., Gu, W., et al.: Dynamic optimal energy flow of integrated electricity and gas systems in continuous space. Appl. Energy **375**, 124052 (2024)
16. Zhang, S., Gu, W., Lu, S., et al.: Integrated energy system analysis-from numerical to analytical (I): analytical method in natural gas system. Proc. CSEE **44**(16), 6432–6443 (2024). (in Chinese)
17. Zhang, S., Gu, W., Lu, S., et al.: Integrated energy system analysis-from numerical to analytical (II): analytical method in district heating system. Proc. CSEE **44**(19), 7603–7615 (2024)
18. Zhang, S., Gu, W., Yu, R., et al.: Modeling and simulation of integrated energy system: review, reflection and prospects. Autom. Electr. Power Syst. **48**(17), 1–21 (2024). (in Chinese)
19. Chen, S., Chen, Q., Xia, Q., Kang, C.: Steady-state security assessment method based on distance to security region boundaries. IET Gener., Transm. Distrib. **7**(3), 288–297 (2013)

20. Wang, Y., You, S., Zhang, H., et al.: Thermal transient prediction of district heating pipeline: optimal selection of the time and spatial steps for fast and accurate calculation. Appl. Energy **206**, 900–910 (2017)
21. Schmidt, M., Aßmann, D., Burlacu, R., et al.: GasLib—A Library of Gas Network Instances (2017)
22. Fang, J., Zeng, Q., Ai, X., Chen, Z., Wen, J.: Dynamic optimal energy flow in the integrated natural gas and electrical power systems. IEEE Trans. Sustain. Energy **9**(1), 188–198 (2018)
23. REFPROP. https://www.nist.gov/programs-projects/reference-fluid-thermodynamic-and-transport-properties-database-refprop

Chapter 5
Resource Aggregation for Low-Carbon Power Systems Under Renewable Uncertainties

5.1 Introduction

One of the notable features of low-carbon power systems is a high penetration of renewable integration. For example, the capacity percentage of renewable installations in China has exceeded 40% by 2024 [1], which will continue a fast development in the future due to the dual carbon goal [2]. However, renewable power is strongly weather-dependent. Consequently, an accurate prediction of renewable integration is still difficult due to the inevitable prediction error on weather [3], evoking the potential risk of energy mismatch in low-carbon power systems. This necessitates the incorporation of renewable uncertainties into resource aggregation for coordinated energy management in low-carbon power systems. Driven by this motivation, this chapter further introduces the renewable uncertainties to the previous resource aggregation methods in Chaps. 2–4 where renewable integration is predetermined without uncertainty considerations.

The remaining part of this chapter is organized below. Section 5.2 will introduce two common representation forms of renewable uncertainties, which are *probabilistic distribution functions* (PDFs) and intervals. Particularly, the PDF form delineates the probabilities of a renewable power prediction, while the interval form describes the upper and lower bounds of a renewable power prediction under predefined confidence levels. Since there are two different forms to represent renewable uncertainties, Sects. 5.3 and 5.4 will accordingly introduce their resource aggregation methods by taking transmission-level resource aggregation as a typical example. Finally, the summary of this chapter will be provided in Sect. 5.5.

W. Lin et al., *Coordinated Energy Management for Low-Carbon Power Systems*,
https://doi.org/10.1007/978-981-96-7533-3_5

5.2 Representations of Renewable Uncertainties

5.2.1 PDF-Based Representation

The PDF-based representation delineates renewable uncertainties by using the PDF of renewable power prediction. Such a representation is generally obtained by using pre-defined types of PDFs (e.g., the Gaussian distribution [4], and the Weibull distribution [5]). For illustration purposes, a PDF-based representation is stated as $f\left(\boldsymbol{\theta}_y, y\right)$, where a renewable power prediction y is a stochastic variable and $\boldsymbol{\theta}_y$ are the parameters to delineate the specific type of the PDF f. It should be noted that the parameters $\boldsymbol{\theta}_y$ may be affected by the factors which influence renewable power, such as weather information (e.g., wind and temperature) and geography information (e.g., altitude and terrain) [6]. Such factors are observed for a renewable power prediction, and denoted as $\mathbf{x}$ hereafter. Consequently, it is necessary to incorporate such influence factors into the PDF f through the parameters $\boldsymbol{\theta}_y$. This can be achieved through a *maximum likelihood estimation* (MLE) problem [7], as elaborated below.

An MLE problem is the following equations associated with the parameters $\boldsymbol{\theta}_y$ and one historical observation set $\mathbf{Y} = \{y_1, y_2, \ldots, y_{\mathrm{N}}\}$:

$$\nabla_{\boldsymbol{\theta}_y} \log f\left(\boldsymbol{\theta}_y, y_i\right) = \mathbf{0}, i = 1, 2, \ldots, \mathrm{N}, \tag{5.1}$$

where $\boldsymbol{\theta}_y$ are regarded as variables in (5.1).

It is not hard to know that the MLE problem is equivalent to solving the following unconstrained optimization problem:

$$\min_{\boldsymbol{\theta}_y} \sum_{i=1}^{\mathrm{N}} \mathbf{1}^\top \left(\nabla_{\boldsymbol{\theta}_y} \log f\left(\boldsymbol{\theta}_y, y_i\right)\right)^2. \tag{5.2}$$

To incorporate the influence factors $\mathbf{x}$ into $\boldsymbol{\theta}_y$, let's employ one surrogate model ϕ which maps $\mathbf{x}$ to $\boldsymbol{\theta}_y$, i.e., $\boldsymbol{\theta}_y = \phi(\mathbf{x})$. Substituting the surrogate model ϕ into (5.2) yields:

$$\min_{\phi} \sum_{i=1}^{\mathrm{N}} \mathbf{1}^\top \left(\nabla_{\boldsymbol{\theta}_y} \log f\left(\boldsymbol{\theta}_y, y_i\right) |_{\boldsymbol{\theta}_y = \boldsymbol{\theta}_y^i}\right)^2, \tag{5.3}$$

$$\text{s.t. } \boldsymbol{\theta}_y^i = \phi\left(\mathbf{x}_i\right), \tag{5.4}$$

where $\mathbf{x}_i$ are the influence factors associated with the historical observation y_i.

When influence factors $\mathbf{x}$ are given, the parameters $\boldsymbol{\theta}_y$ can be known as $\boldsymbol{\theta}_y = \phi^*(\mathbf{x})$ where ϕ^* is the solutions from the optimization problem (5.3)–(5.4). In such a manner, the PDF-based representation can be got as $f\left(\phi^*(\mathbf{x}), y\right)$ when the influence factors are given during resource aggregation.

5.2.2 Interval-Based Representation

The interval-based representation delineates renewable uncertainties by using the upper and lower bounds of the renewable power prediction y when the influence factors $\mathbf{x}$ are given. Particularly, the bounds are selected as quantile values under certain quantile levels for the stochastic variable y. Particularly, the quantile value $Q_y(\tau)$ associated with the quantile level $\tau \in (0, 1)$ of the stochastic variable y is defined as

$$Q_y(\tau) = \inf\{y : F(y) \geq \tau\}, \tag{5.5}$$

where the function F is the *cumulative distribution function* (CDF) of the stochastic variable y. A quantile value is mathematically the lower bound of y when its CDF value is larger than or equal to the quantile level τ. For example, $\left(Q_y\left(\tau = 0.7\right) = 90\right)$ means that the probability of the renewable power prediction being less than or equal to 90 is 70%.

In this context, the key of the interval-based representation is to compute the quantile value associated with one quantile level when influence factors $\mathbf{x}$ are given. This will be resolved by employing a surrogate model ψ which maps $(\mathbf{x}, \tau)$ to the quantile value, i.e., $\hat{y}^\tau = \psi\left(\mathbf{x}, \tau\right)$. This task can be accomplished by using the following Pinball loss [8]:

$$L_\tau\left(y, \hat{y}^\tau\right) = \begin{cases} (1-\tau)\left(\hat{y}^\tau - y\right) & \hat{y}^\tau \geq y \\ \tau\left(y - \hat{y}^\tau\right) & \hat{y}^\tau < y \end{cases}, \tag{5.6}$$

By considering $y^\tau - y$ as the horizontal axis and $L_\tau\left(y, \hat{y}^\tau\right)$ as the vertical axis, the graph of the Pinball loss (5.6) is plotted in Fig. 5.1. It can be found in Fig. 5.1 that minimizing the Pinball loss (5.6) contributes to minimizing the deviation between the true quantile value y^τ and its estimation through $\hat{y}^\tau = \psi\left(\mathbf{x}, \tau\right)$ under the quantile level τ. Consequently, the surrogate model $\hat{y}^\tau = \phi\left(\mathbf{x}, \tau\right)$ can be determined by minimizing the Pinball loss, i.e.,

$$\phi^*(\tau) = \arg\min_{\phi} L_\tau\left(\hat{y}^\tau, y\right). \tag{5.7}$$

When influence factors $\mathbf{x}$ and the quantile level τ is given, the quantile value $\hat{y}^\tau = \psi\left(\mathbf{x}, \tau\right)$ can be known as either the upper bound or the lower bound of a renewable power prediction to reflect renewable uncertainties.

5.2.3 Selections of Surrogate Models in Representations

The surrogate models ϕ and ψ are required in Sects. 5.2.1 and 5.2.2 to represent renewable uncertainties based on the influence factors $\mathbf{x}$. Such surrogate models may be selected from many candidate models (e.g., a deep neural network, or a

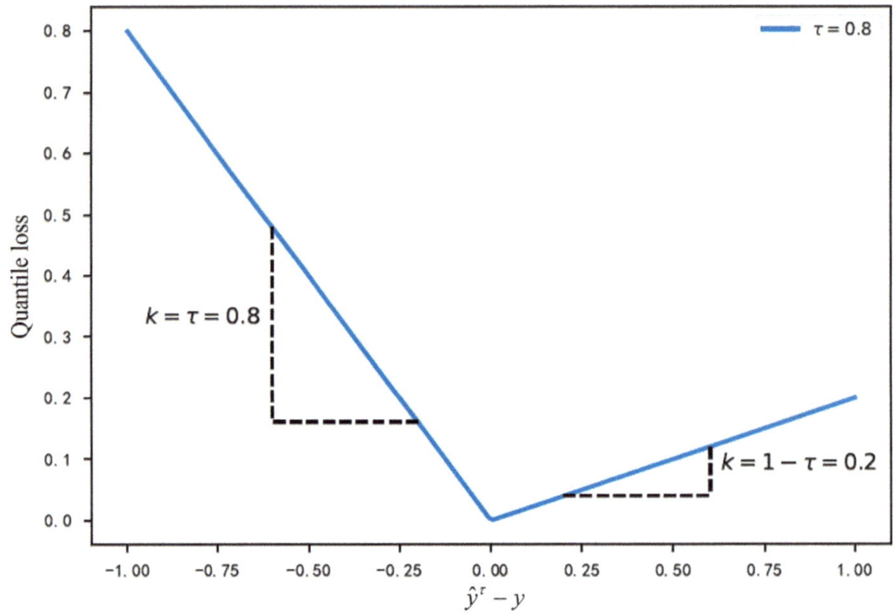

Fig. 5.1 Illustration of the Pinball loss ($\tau = 0.8$)

linear regression model). Consequently, one remaining task for representations of renewable uncertainties is to select desirable surrogate models. This task is suggested in this chapter by comparing the following index under different candidate models:

$$\mathrm{UA} = \frac{1}{\mathrm{N}} \sum_{\tau=\tau_1}^{\tau_\mathrm{N}} L_\tau \left(y, \hat{y}^\tau\right), \tag{5.8}$$

The index UA (5.8) comes from the current industry standard [9] which puts forward accuracy requirements on renewable power predictions. The index UA (5.8) computes the deviation of different quantile levels under a given surrogate model. Consequently, its smaller value indicates a higher prediction accuracy, which means reduced renewable uncertainties from the given surrogate model. In this manner, candidate models with higher prediction accuracy (i.e., higher values of the index UA (5.8)) can be selected to reduce the impacts of representing renewable uncertainties.

5.3 Chance Constrains-Based Resource Aggregation

5.3.1 Problem Formulation

Each transmission sub-system is interconnected with other transmission sub-systems through tie-line power flow exchanges $\mathbf{P}_{\mathrm{B}}$. The constraints of one transmission sub-system can be categorized into three types: (1) power flow constraints; (2) chance constraints; and (3) bounds of variables. These three types of constraints are described below.

(1) Power balance constraints

The power injected into one node should equal its power flowing out, i.e.,

$$\boldsymbol{e}_{\mathrm{G}}^{\top}\boldsymbol{P}_{\mathrm{G}}+\boldsymbol{e}_{\mathrm{B}}^{\top}\boldsymbol{P}_{\mathrm{B}}=\boldsymbol{e}_{\mathrm{D}}^{\top}\boldsymbol{P}_{\mathrm{D}} \tag{5.9}$$

where $\mathbf{P}_{\mathrm{G}}$, $\mathbf{P}_{\mathrm{B}}$, and $\mathbf{P}_{\mathrm{D}}$ are active generation levels, tie-line power flow exchanges, and net power demands, respectively; $\mathbf{e}_{\mathrm{G}}$, $\mathbf{e}_{\mathrm{B}}$, and $\mathbf{e}_{\mathrm{D}}$ are all-one row vectors associated with $\mathbf{P}_{\mathrm{G}}$, $\mathbf{P}_{\mathrm{B}}$, and $\mathbf{P}_{\mathrm{D}}$, respectively; $\mathbf{P}_{\mathrm{D}}$ is stochastic variables which require uncertainty representations in Sect. 5.2.

(2) Chance constraints

Branch flows should not exceed the limitations given confidence levels, i.e.,

$$\Pr\left(\mathbf{P}_{\mathrm{f}}\leq\overline{\mathbf{P}}_{\mathrm{f}}\right)\geq\boldsymbol{\varepsilon}_{\overline{\mathrm{f}}}, \tag{5.10}$$

$$\Pr\left(\underline{\mathbf{P}}_{\mathrm{f}}\leq\mathbf{P}_{\mathrm{f}}\right)\geq\boldsymbol{\varepsilon}_{\underline{\mathrm{f}}}, \tag{5.11}$$

$$\mathbf{P}_{\mathrm{f}}=\mathbf{S}\left(\mathbf{M}_{\mathrm{G}}\mathbf{P}_{\mathrm{G}}+\mathbf{M}_{\mathrm{B}}\mathbf{P}_{\mathrm{B}}+\mathbf{M}_{\mathrm{D}}\mathbf{P}_{\mathrm{D}}\right), \tag{5.12}$$

where $\underline{\mathbf{P}}_{\mathrm{f}}$ and $\overline{\mathbf{P}}_{\mathrm{f}}$ are the lower and upper bounds of $\mathbf{P}_{\mathrm{f}}$, respectively; $\Pr(\mathbf{X}\leq\mathbf{Y})\geq\mathbf{Z}$ indicates that a probability where the stochastic variable x_i is not larger than the value y_i will not be less than the value z_i, where x_i, y_i and z_i are the ith elements in $\mathbf{X}$, $\mathbf{Y}$ and $\mathbf{Z}$, respectively; $\boldsymbol{\varepsilon}_{\overline{\mathrm{f}}}$ and $\boldsymbol{\varepsilon}_{\underline{\mathrm{f}}}$ are the confidence levels of fulfilling the upper bounds and lower bounds, respectively; $\mathbf{S}$ is the *power transfer distribution factor* (PTDF) matrix.

(3) Bounds of variables

(1) Generation levels $\mathbf{P}_{\mathrm{G}}$ should not exceed their lower bounds $\underline{\mathbf{P}}_{\mathrm{G}}$ and upper bounds $\overline{\mathbf{P}}_{\mathrm{G}}$, i.e.,

$$\underline{\boldsymbol{P}}_{\mathrm{G}}\leq\boldsymbol{P}_{\mathrm{G}}\leq\overline{\boldsymbol{P}}_{\mathrm{G}}. \tag{5.13}$$

(2) Tie-line power flow exchanges $\mathbf{P}_{\mathrm{B}}$ should not exceed their lower bounds $\underline{\mathbf{P}}_{\mathrm{B}}$ and upper bounds $\overline{\mathbf{P}}_{\mathrm{B}}$, i.e.,

$$\underline{\mathbf{P}}_{\mathrm{B}} \leq \mathbf{P}_{\mathrm{B}} \leq \overline{\mathbf{P}}_{\mathrm{B}}. \tag{5.14}$$

For ease of discussions, the constraints (5.9)–(5.14) are compared as

$$\mathbf{h}_1\mathbf{P}_{\mathrm{G}} + \mathbf{h}_2\mathbf{P}_{\mathrm{B}} + \mathbf{h}_3\mathbf{P}_{\mathrm{D}} = \mathbf{h}_4, \tag{5.15}$$

$$\Pr\left(\mathbf{g}_1\mathbf{P}_{\mathrm{G}} + \mathbf{g}_2\mathbf{P}_{\mathrm{B}} + \mathbf{b} \leq \mathbf{g}_3\mathbf{P}_{\mathrm{D}}\right) \geq \boldsymbol{\varepsilon}, \tag{5.16}$$

$$\underline{\mathbf{P}}_{\mathrm{G}} \leq \mathbf{P}_{\mathrm{G}} \leq \overline{\mathbf{P}}_{\mathrm{G}}, \tag{5.17}$$

$$\underline{\mathbf{P}}_{\mathrm{B}} \leq \mathbf{P}_{\mathrm{B}} \leq \overline{\mathbf{P}}_{\mathrm{B}} \tag{5.18}$$

where the constants $\mathbf{h}_1$, $\mathbf{h}_2$, $\mathbf{h}_3$, and $\mathbf{h}_4$ are from the constraint (5.9); the constants $\mathbf{g}_1$, $\mathbf{g}_2$, $\mathbf{g}_3$, $\mathbf{b}$ and the confidence levels $\boldsymbol{\varepsilon}$ are from (5.10)–(5.12). Note that smaller confidence levels $\boldsymbol{\varepsilon}$ indicate the smaller probability of constraint violations. Consequently, the confidence levels can be chosen based on the tolerance of constraint violations. Particularly, the stochastic variables $\mathbf{P}_{\mathrm{D}}$ in the constraints (5.15)–(5.18) are delineated by PDFs in Sect. 5.2.1.

The resource aggregation at the transmission level when facing renewable uncertainties delineated by PDFs is to find the following permissible region of tie-line power flow exchanges:

$$\mathbf{W}_{\mathrm{PDF}} \triangleq \{\mathbf{P}_{\mathrm{B}} \mid \exists \mathbf{P}_{\mathrm{G}} \text{ fulfilling } (5.15)-(5.18)\}. \tag{5.19}$$

5.3.2 *Resource Aggregation Based on Conversion of Chance Constraints*

When the transmission-level constraints are linear, Chap. 2 points out that the multi-parametric programming theory in [10] is ideal for computing one permissible region of tie-line power flow exchanges. However, the chance constraints (5.16) are nonlinear; thus, the multi-parametric programming theory cannot be directly applied to the constraints (5.15)–(5.18). To resolve this issue which prevents the employment of the multi-parametric programming theory towards one permissible region of tie-line power flow exchanges, the task comes down to formulating the chance constraints in a linear form. Such a formulation will be introduced in Sect. 5.3.2 by developing two states. The first stage is to convert all the constraints associated with the stochastic variables $\mathbf{P}_{\mathrm{D}}$ to chance constraints, and the second stage is to equivalently formulate the resultant chance constraints by linear constraints using quantile values. These two stages are sequentially introduced below.

5.3.2.1 First Stage for Chance Constraint Conversion

Although the constraints (5.17) are not expressed as chance constraints, they are also affected by stochastic variables $\mathbf{P}_{\mathrm{D}}$ through the constraint (5.15). Consequently, this stage eliminates the constraint (5.15) based on Gaussian elimination so that the constraints (5.17) can be also converted to chance constraints.

Power balance should hold even if renewable uncertainties exist. The generators will be adjusted to eliminate the unbalanced power arising from renewable uncertainties. Different adjustment rules can be designed and then incorporated into the elimination process. For the sake of simplification, a fraction of generators (called"adjustable generators") will be assumed to adjust their generation levels to eliminate the unbalance arising from renewable uncertainties, while other generators (called "fixed generators") fix their generation levels. This hypothesis stems from the conventional power analysis, where most generators fix their generation levels while only one generator or a fraction of generators can adjust generation levels to eliminate the unbalance [11].

Suppose that only one generator is allowed to adjust its generation levels to ensure the constraint (5.15) under renewable uncertainties. Consequently, the constraint (5.15) can be reformulated in a separable version as

$$\mathbf{h}_{1\mathrm{A}}\mathbf{P}_{\mathrm{GA}} + \mathbf{h}_{1\mathrm{F}}\mathbf{P}_{\mathrm{GF}} + \mathbf{h}_{2}\mathbf{P}_{\mathrm{B}} + \mathbf{h}_{3}\mathbf{P}_{\mathrm{D}} = \mathbf{h}_{4}, \tag{5.20}$$

where $\mathbf{P}_{\mathrm{GA}}$ and $\mathbf{P}_{\mathrm{GF}}$ are generation levels associated with adjustable generators and fixed generators, respectively; $\mathbf{h}_1 = \left[\mathbf{h}_{1\mathrm{A}}\ \mathbf{h}_{1\mathrm{F}}\right]$; $\mathbf{h}_{1\mathrm{A}}$ and $\mathbf{h}_{1\mathrm{F}}$ are sub-matrices associated with adjustable generators and fixed generators, respectively.

Based on the version in (5.20), the generation levels $\mathbf{P}_{\mathrm{GA}}$ can be formulated as a linear function of the stochastic variables $\mathbf{P}_{\mathrm{D}}$, i.e.,

$$\begin{aligned}\mathbf{P}_{\mathrm{GA}} &= -\mathbf{h}_{1\mathrm{A}}^{-1}\mathbf{h}_{1\mathrm{F}}\mathbf{P}_{\mathrm{GF}} - \mathbf{h}_{1\mathrm{A}}^{-1}\mathbf{h}_{2}\mathbf{P}_{\mathrm{B}} - \mathbf{h}_{1\mathrm{A}}^{-1}\mathbf{h}_{3}\mathbf{P}_{\mathrm{D}} + \mathbf{h}_{1A}^{-1}\mathbf{h}_{4} \\ &= \mathbf{H}_{1}\mathbf{P}_{\mathrm{GF}} + \mathbf{H}_{2}\mathbf{P}_{\mathrm{B}} + \mathbf{H}_{3}\mathbf{P}_{\mathrm{D}} + \mathbf{H}_{4}\end{aligned}. \tag{5.21}$$

The constraints (5.17) associated with $\mathbf{P}_{\mathrm{GA}}$ indicate the limitations of generator capacities. Considering the renewable uncertainties in $\mathbf{P}_{\mathrm{GA}}$ from (5.21), $\mathbf{P}_{\mathrm{GA}}$ should satisfy their limitations at the given confidence levels $\boldsymbol{\varepsilon}_{\underline{\mathrm{GA}}}$ and $\boldsymbol{\varepsilon}_{\overline{\mathrm{GA}}}$, i.e.,

$$\Pr\left(\mathbf{H}_{3}\mathbf{P}_{\mathrm{D}} \le \overline{\mathbf{P}}_{\mathrm{GA}} - \mathbf{H}_{1}\mathbf{P}_{\mathrm{GF}} - \mathbf{H}_{2}\mathbf{P}_{\mathrm{B}} - \mathbf{H}_{4}\right) \ge \boldsymbol{\varepsilon}_{\overline{\mathrm{GA}}}, \tag{5.22}$$

$$\Pr\left(-\mathbf{H}_{3}\mathbf{P}_{R} \le -\underline{\mathbf{P}}_{\mathrm{GA}} + \mathbf{H}_{1}\mathbf{P}_{\mathrm{GF}} + \mathbf{H}_{2}\mathbf{P}_{\mathrm{B}} + \mathbf{H}_{4}\right) \ge \boldsymbol{\varepsilon}_{\underline{\mathrm{GA}}}, \tag{5.23}$$

where $\overline{\mathbf{P}}_{\mathrm{GA}}$ and $\underline{\mathbf{P}}_{\mathrm{GA}}$ are the upper and lower bounds of adjustable generators from (5.17).

Additionally, substituting (5.21) into the original chance constraints (5.16) yields the following new version of the chance constraints (5.16):

$$\Pr\left(\mathbf{G}_1\mathbf{P}_{\mathrm{GF}} + \mathbf{G}_2\mathbf{P}_{\mathrm{B}} + \mathbf{G}_4 \le \mathbf{G}_3\mathbf{P}_{\mathrm{D}}\right) \ge \boldsymbol{\varepsilon}, \tag{5.24}$$

where $\mathbf{G}_1 = \mathbf{g}_{1\mathrm{F}} + \mathbf{g}_{1\mathrm{A}}\mathbf{H}_1$; $\mathbf{g}_1 = \left[\mathbf{g}_{1\mathrm{A}}\ \mathbf{g}_{1\mathrm{F}}\right]$; $\mathbf{g}_{1\mathrm{A}}$ and $\mathbf{g}_{1\mathrm{F}}$ are the sub-matrices associated with adjustable generators and fixed generators, respectively; $\mathbf{G}_2 = \mathbf{g}_{1\mathrm{A}}\mathbf{H}_2 + \mathbf{g}_2$; $\mathbf{G}_3 = \mathbf{g}_3 - \mathbf{g}_{1\mathrm{A}}\mathbf{H}_3$; $\mathbf{G}_4 = \mathbf{g}_{1\mathrm{A}}\mathbf{H}_4 + \mathbf{b}$.

Consequently, the constraints (5.15)–(5.18) can be equivalently converted as

$$\underline{\mathbf{P}}_{\mathrm{GF}} \le \mathbf{P}_{\mathrm{GF}} \le \overline{\mathbf{P}}_{\mathrm{GF}}, \tag{5.25}$$

$$\underline{\mathbf{P}}_{\mathrm{B}} \le \mathbf{P}_{\mathrm{B}} \le \overline{\mathbf{P}}_{\mathrm{B}}, \tag{5.26}$$

$$\Pr\left(-\mathbf{F}_1\mathbf{P}_{\mathrm{GF}} - \mathbf{F}_2\mathbf{P}_{\mathrm{B}} - \mathbf{F}_4 \ge -\mathbf{F}_3\mathbf{P}_{\mathrm{D}}\right) \ge \boldsymbol{\varepsilon}_{\mathrm{new}}, \tag{5.27}$$

where the constant matrices $\mathbf{F}_1$, $\mathbf{F}_2$ and $\mathbf{F}_3$ and the constant vector $\mathbf{F}_4$ are from the constraints (5.22)–(5.24); $\boldsymbol{\varepsilon}_{\mathrm{new}}$ compact the confidence levels in the constraints (5.22)–(5.24).

5.3.2.2 Second Stage for Chance Constraint Conversion

This stage reformulates the chance constraints (5.27) in a linear form by using quantile values of stochastic variables. Suppose the ith chance constraint in the chance constraints (5.27) can be given by

$$\Pr\left(-\mathbf{F}_1^{(i,\#)}\mathbf{P}_{\mathrm{GF}} - \mathbf{F}_2^{(i,\#)}\mathbf{P}_{\mathrm{B}} - \mathbf{F}_4^{(i,\#)} \ge -\mathbf{F}_3^{(i,\#)}\mathbf{P}_{\mathrm{D}}\right) \ge \varepsilon_{\mathrm{new}}^{i}, \tag{5.28}$$

where $*^i$ indicates the ith element of the variables $*$; $*^{(i,\#)}$ indicates a matrix formed by the ith row of $*$.

By regarding $-\mathbf{F}_3^{(i,\#)}\mathbf{P}_{\mathrm{D}}$ as new stochastic variables $\boldsymbol{\eta}^i$, the corresponding CDF $\zeta_{\eta i}$ paves one opportunity to formulate the chance constraint (5.28) in a linear form, as explained in Fig. 5.2.

When the value of $-\mathbf{F}_{1\,\mathrm{n}}^{(i,\#)}\mathbf{P}_{\mathrm{GF}} - \mathbf{F}_2^{(i,\#)}\mathbf{P}_{\mathrm{B}} - \mathbf{F}_4^{(i,\#)}$ is larger than the quantile value $X_{\eta i}$, the chance constraint (5.28) is fulfilled. Consequently, the chance constraint (5.28) can be reformulated as a linear constraint by employing the quantile value $X_{\eta i}$:

$$-\mathbf{F}_1^{(i,\#)}\mathbf{P}_{\mathrm{GF}} - \mathbf{F}_2^{(i,\#)}\mathbf{P}_{\mathrm{B}} - \mathbf{F}_4^{(i,\#)} \ge X_{\eta i}. \tag{5.29}$$

It is not hard to know that the key to constructing the linear constraint (5.29) is to capture the CDF $\zeta_{\eta i}$. To accomplish this task, the strategy based on a Cholesky decomposition [13] will be introduced below.

The stochastic variables $\mathbf{P}_{\mathrm{D}}$ are assumed to follow the same distributions, whose expected values and variances are denoted as $\boldsymbol{\mu}_{\mathrm{D}}$ and $\boldsymbol{\sigma}_{\mathrm{D}}^2$, respectively. The correlations among $\mathbf{P}_{\mathrm{D}}$ can be represented using a Pearson correlation matrix $\boldsymbol{\rho}_{\mathrm{D}}$.

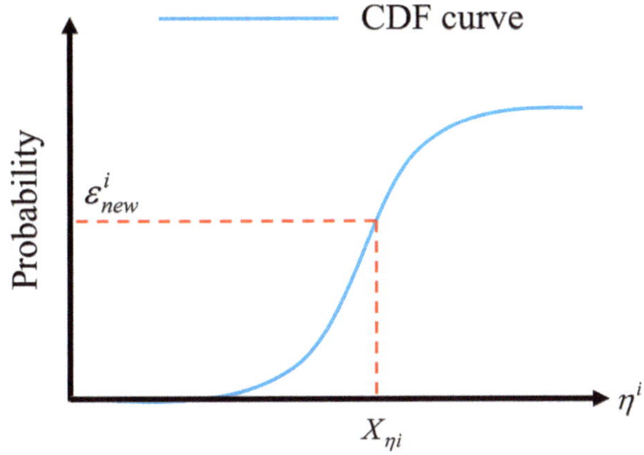

Fig. 5.2 Quantile in a CDF curve, reprinted from [12], open access

The stochastic variables $\mathbf{P}_{\mathrm{D}}$ with the correlation matrix $\boldsymbol{\rho}_{\mathrm{D}}$ can be transformed into independent variables $\boldsymbol{\gamma}$ based on a Cholesky decomposition as follows:

$$\mathbf{P}_{\mathrm{D}} = \mathbf{G}\boldsymbol{\gamma}, \tag{5.30}$$

where $\mathbf{G}$ is obtained by $\boldsymbol{\rho} = \mathbf{G}\mathbf{G}^{\top}$ where $\mathbf{G}$ is a triangular decomposition of $\boldsymbol{\rho}$. This transformation is detailed in Appendix 5.6.1.

Additionally, the new stochastic variables $\boldsymbol{\eta}^{i} = -\mathbf{F}_{3}^{(i,\#)}\mathbf{P}_{\mathrm{D}}$ can be formulated as a function of $\boldsymbol{\gamma}$ considering (5.30), i.e.,

$$\boldsymbol{\eta} = -\mathbf{G}\mathbf{F}_{3}^{(i,\#)}\boldsymbol{\gamma} = \mathbf{G}_{\eta}\boldsymbol{\gamma}. \tag{5.31}$$

Based on (5.31), the new stochastic variables $\boldsymbol{\eta}^{i}$ are a linear combination of the independent variables $\boldsymbol{\gamma}$ and thus follow the same distributions of the original stochastic variables $\mathbf{P}_{\mathrm{D}}$. If their expected values $\boldsymbol{\mu}_{\eta}$ and variances $\boldsymbol{\sigma}_{\eta}^{2}$ are known, the CDF $\zeta_{\eta i}$ associated with the ith element of $\boldsymbol{\eta}^{i}$ can be directly delineated. Consequently, the next task is to express $\boldsymbol{\mu}_{\eta}$ and $\boldsymbol{\sigma}_{\eta}^{2}$ based on the known $\boldsymbol{\mu}_{\mathrm{D}}$ and $\boldsymbol{\sigma}_{\mathrm{D}}^{2}$.

Based on (5.30), the expected values $\boldsymbol{\mu}_{\gamma}$ and variances $\boldsymbol{\sigma}_{\gamma}^{2}$ associated with $\boldsymbol{\gamma}$ can be calculated using $\boldsymbol{\mu}_{\mathrm{D}}$ and $\boldsymbol{\sigma}_{\mathrm{D}}^{2}$ as follows:

$$\begin{cases} \boldsymbol{\mu}_{\gamma} = \mathbf{G}^{-1}\boldsymbol{\mu}_{\mathrm{D}} \\ \boldsymbol{\sigma}_{\gamma}^{2} = \left(\mathbf{G}^{-1}\right)^{2}\boldsymbol{\sigma}_{\mathrm{D}}^{2} \end{cases}. \tag{5.32}$$

Additionally, the expected values $\boldsymbol{\mu}_{\eta}$ and variances $\boldsymbol{\sigma}_{\eta}^{2}$ associated with $\boldsymbol{\eta}^{i}$ can be calculated based on (5.31) by using $\boldsymbol{\mu}_{\gamma}$ and $\boldsymbol{\sigma}_{\gamma}^{2}$ as follows:

$$\begin{cases} \boldsymbol{\mu}_{\eta} = \mathbf{G}_{\eta}\boldsymbol{\mu}_{\gamma} \\ \boldsymbol{\sigma}_{\eta}^{2} = \left(\mathbf{G}_{\eta}^{-1}\right)^{2}\boldsymbol{\sigma}_{\gamma}^{2} \end{cases}. \tag{5.33}$$

Combining (5.32) and (5.33), $\boldsymbol{\mu}_\eta$ and $\boldsymbol{\sigma}_\eta^2$ can be explicitly computed based on the known $\boldsymbol{\mu}_{\mathrm{D}}$ and $\boldsymbol{\sigma}_{\mathrm{D}}^2$ as

$$\begin{cases} \boldsymbol{\mu}_\eta = \mathbf{G}_\eta \mathbf{G}^{-1} \boldsymbol{\mu}_{\mathrm{D}} \\ \boldsymbol{\sigma}_\eta^2 = \left(\mathbf{G}_\eta^{-1}\right)^2 \left(\mathbf{G}^{-1}\right)^2 \boldsymbol{\sigma}_{\mathrm{D}}^2 \end{cases}. \tag{5.34}$$

Based on the derivations above, the new stochastic variables $\boldsymbol{\eta}^i = -\mathbf{F}_3^{(i,\#)}\mathbf{P}_{\mathrm{D}}$ in the chance constraints follow the same distributions of the stochastic variables $\mathbf{P}_{\mathrm{D}}$, and their expected values and variances can be explicitly formulated as (5.34). After implementing the two stages above, the constraints (5.25)–(5.27) can be converted into the following linear constraints for their employment of the multi-parametric programming theory towards the permissible region of tie-line power flow exchanges under renewable uncertainties:

$$\underline{\mathbf{P}}_{\mathrm{GF}} \leq \mathbf{P}_{\mathrm{GF}} \leq \overline{\mathbf{P}}_{\mathrm{GF}}, \tag{5.35}$$

$$\underline{\mathbf{P}}_{\mathrm{B}} \leq \mathbf{P}_{\mathrm{B}} \leq \overline{\mathbf{P}}_{\mathrm{B}}, \tag{5.36}$$

$$-\mathbf{F}_1\mathbf{P}_{\mathrm{GF}} - \mathbf{F}_2\mathbf{P}_{\mathrm{B}} - \mathbf{F}_4 \geq \mathbf{X}_\eta, \tag{5.37}$$

where $\mathbf{X}_\eta$ stacks all the quantile values from each CDF $\zeta_{\eta i}$.

5.3.3 *Case Studies*

The resource aggregation method in Sect. 5.3.2 will be examined in the following IEEE 9-bus and IEEE 118-bus test systems [14]:

- IEEE 9-bus test system: Two tie-lines are connected at border bus 5 and border bus 9. Two renewable power plants are located at border bus 4 and border bus 7. The capacity of each renewable power plant is 0.3 p.u.
- IEEE 118-bus test system: Three tie lines are connected at border bus 5, border bus 9, and border bus 60. Four renewable power plants are located at border bus 20, border bus 30, border bus 40, and border bus 50. The capacity of each renewable power plant is 3.0 p.u.

In the coming numerical experiments, P_i denotes the tie-line power flow exchanges at border bus i.

5.3.3.1 Effectiveness of the Resource Aggregation Method in Sect. 5.3.2

The IEEE 9-bus test system will be utilized to demonstrate the permissible region of tie-line power flow exchanges under renewable uncertainties delineated by PDFs. When renewable uncertainties delineated by PDFs are incorporated into transmission-level constraints by chance constraints, such a permissible region is given in Fig. 5.3.

To check the validation of the permissible region in Fig. 5.3, the Monte Carlo framework is used. When renewable uncertainties are described by PDFs (normal distributions are employed in the numerical experiments), 5×10^3 random points $\left(P_5^j, P_9^j\right)$ $(j = 1, 2, \ldots, 5 \times 10^3)$ that fall into the permissible region are generated. For a given point, the fulfillment of chance constraints (5.16) is checked by using 5×10^3 renewable samples generated based on PDFs. The numerical experiments show that all the generated points can satisfy the chance constraints (5.16) under predefined confidence levels. This validates the effectiveness of the resource aggregation method in Sect. 5.3.2, which is able to correctly identify the permissible region of tie-line power flow exchanges under renewable uncertainties delineated by PDFs.

5.3.3.2 Investigation of Confidence Levels in Chance Constraints

When renewable uncertainties are delineated by PDFs, the confidence levels $\boldsymbol{\varepsilon}$ in (5.16) indicate the tolerance on the possibility of constraint violations. The higher $\boldsymbol{\varepsilon}$ represent a lower tolerance of constraint violations when facing renewable uncertainties. Section 5.3.3.2 will employ the IEEE 118-bus test system to investigate the impacts of confidence levels $\boldsymbol{\varepsilon}$ on the permissible region of tie-line power flow exchanges. When the confidence levels $\boldsymbol{\varepsilon}$ are set to 0.1, 0.5, and 0.9, the resultant permissible regions are identified. For ease of comparison, their permissible regions are visualized in the domain of (P_5, P_9) in Fig. 5.4.

As shown in Fig. 5.4, the permissible region of tie-line power flow exchanges shrinks with increasing confidence levels $\boldsymbol{\varepsilon}$. This is because the tolerance of constraint violations decreases when the confidence levels $\boldsymbol{\varepsilon}$ increase. Certain combinations of

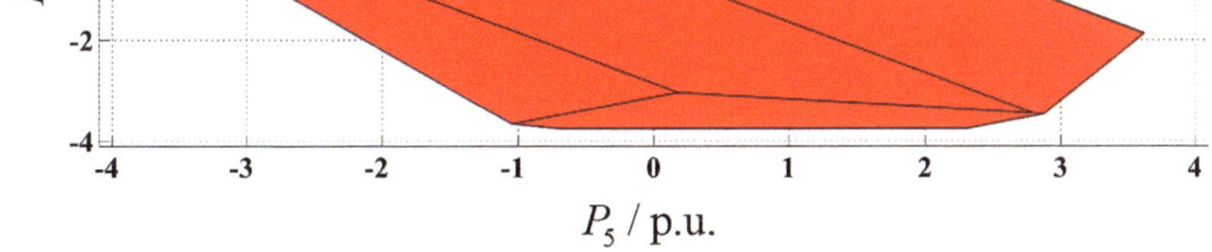

Fig. 5.3 Permissible region of tie-line power flow exchanges under renewable uncertainties delineated by PDFs (IEEE 9-bus test system), reprinted from [12], open access

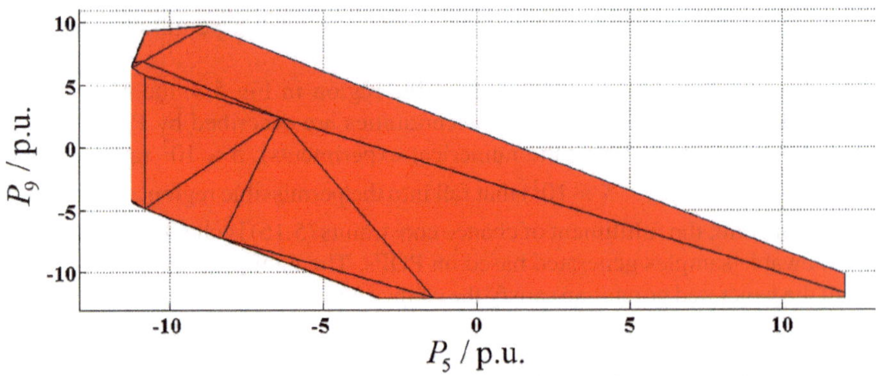

(a) Permissible region in the domain of (P_5, P_9) when $\varepsilon = 0.1$

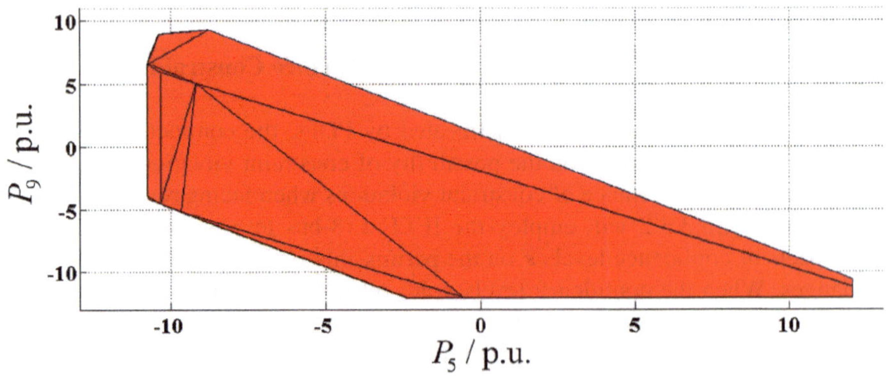

(b) Permissible region in the domain of (P_5, P_9) when $\varepsilon = 0.5$

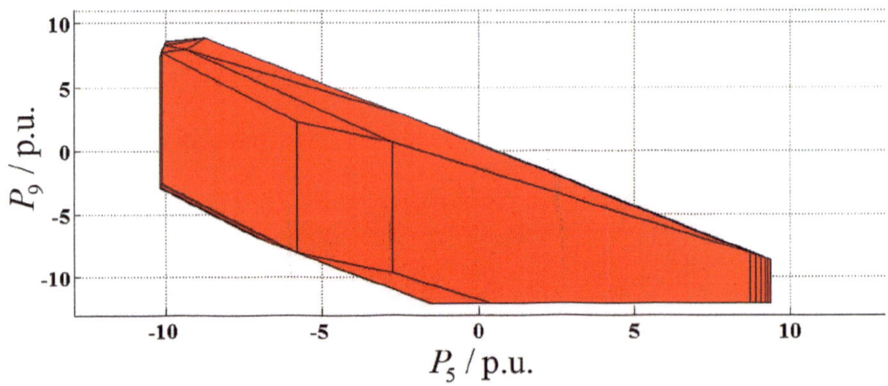

(c) Permissible region in the domain of (P_5, P_9) when $\varepsilon = 0.9$

Fig. 5.4 Permissible regions in the domain of (P_5, P_9) under different confidence levels (IEEE 118-bus test system), reprinted from [12], open access

tie-line power flow exchanges which satisfy the constraints at lower confidence levels will not fulfill the constraints at higher confidence levels. Additionally, fewer combinations of tie-line power flow exchanges obviously lead to smaller permissible regions. Consequently, the permissible region of tie-line power flow exchanges shrinks as the confidence levels $\boldsymbol{\varepsilon}$ increase.

5.4 Scenario-Based Resource Aggregation

5.4.1 Problem Formulation

Similar to the problem formulation in Sect. 5.3.1, the following constraints of one transmission sub-system can be formulated when renewable uncertainties are described by intervals $\left[\mathbf{P}_{\mathrm{D}}^{-}, \mathbf{P}_{\mathrm{D}}^{+}\right]$:

$$\mathbf{h}_1\mathbf{P}_{\mathrm{G}} + \mathbf{h}_2\mathbf{P}_{\mathrm{B}} + \mathbf{h}_3\mathbf{P}_{\mathrm{D}} = \mathbf{h}_4, \tag{5.38}$$

$$\mathbf{g}_1\mathbf{P}_{\mathrm{G}} + \mathbf{g}_2\mathbf{P}_{\mathrm{B}} + \mathbf{b} \leq \mathbf{g}_3\mathbf{P}_{\mathrm{D}}, \tag{5.39}$$

$$\underline{\mathbf{P}}_{\mathrm{G}} \leq \mathbf{P}_{\mathrm{G}} \leq \overline{\mathbf{P}}_{\mathrm{G}}, \tag{5.40}$$

$$\underline{\mathbf{P}}_{\mathrm{B}} \leq \mathbf{P}_{\mathrm{B}} \leq \overline{\mathbf{P}}_{\mathrm{B}}, \tag{5.41}$$

$$\forall \mathbf{P}_{\mathrm{D}} \in \left[\mathbf{P}_{\mathrm{D}}^{-}, \mathbf{P}_{\mathrm{D}}^{+}\right], \tag{5.42}$$

where the constant matrices $\mathbf{h}_1$, $\mathbf{h}_2$, $\mathbf{h}_3$, $\mathbf{g}_1$, $\mathbf{g}_2$, $\mathbf{g}_3$ and the constant vectors $\mathbf{h}_4$ and $\mathbf{g}_3$ can be obtained from the constraints (5.15)–(5.18); $\mathbf{P}_{\mathrm{D}}^{-}$ and $\mathbf{P}_{\mathrm{D}}^{+}$ are the lower and upper bounds of the stochastic variables $\mathbf{P}_{\mathrm{D}}$, respectively. Note that a more accurate prediction technique can lead to a smaller length of the interval $\left[\mathbf{P}_{\mathrm{D}}^{-}, \mathbf{P}_{\mathrm{D}}^{+}\right]$.

The resource aggregation at the transmission level when facing renewable uncertainties delineated by intervals is to find the following permissible region of tie-line power flow exchanges:

$$\mathbf{W}_{\text{Interval}} \triangleq \{\mathbf{P}_{\mathrm{B}} \mid \exists \mathbf{P}_{\mathrm{G}} \text{ fulfilling } (5.38) - (5.42)\}. \tag{5.43}$$

5.4.2 Resource Aggregation Based on Extreme Scenarios

As introduced in Chap. 2, the multi-parametric programming theory in [10] is ideal for computing one permissible region of tie-line power flow exchanges when facing finite net power demand scenarios. However, the interval constraints (5.42) indicate infinite net power demand scenarios for computing the permissible region of tie-line power flow exchanges (5.43). To resolve this conflict, Sect. 5.4.2.1 will introduce linear constraints with finite scenarios, by which the permissible region of tie-line power flow exchanges (5.43) can still be computed. In addition, a scenario-based decomposition strategy will be introduced in Sect. 5.4.2.2 for its accelerated computation.

5.4.2.1 Linear Constraints with Finite Scenarios

Let's consider the following linear constraints with finite net power demand scenarios:

$$\mathbf{h}_1\mathbf{P}_{\mathrm{G}} + \mathbf{h}_2\mathbf{P}_{\mathrm{B}} + \mathbf{h}_3\mathbf{P}_{\mathrm{D}}^s = \mathbf{h}_4, s \in \mathbf{S}, \tag{5.44}$$

$$\mathbf{g}_1\mathbf{P}_{\mathrm{G}}^s + \mathbf{g}_2\mathbf{P}_{\mathrm{B}} + \mathbf{b} \leq \mathbf{g}_3\mathbf{P}_{\mathrm{D}}^s, s \in \mathbf{S}, \tag{5.45}$$

$$\text{Constraints (5.40)-(5.41)}, \tag{5.46}$$

where $\mathbf{P}_{\mathrm{D}}^s$ are the net power demands under the sth extreme scenario, and the net power demands $P_{\mathrm{D},i}$ in $\mathbf{P}_{\mathrm{D}}$ equals its either lower bound $P_{\mathrm{D},i}^-$ or upper bound $P_{\mathrm{D},i}^+$ in each extreme scenario; the set $\mathbf{S}$ is composed of all the possible extreme scenarios.

Based on the constraints (5.44)–(5.46), one permissible region of tie-line power flow exchanges can be defined below.

$$\widetilde{\mathbf{W}}_{\mathrm{Interval}} \triangleq \{\mathbf{P}_{\mathrm{B}} \mid \exists \mathbf{P}_{\mathrm{G}} \text{ fulfilling } (5.44) - (5.46)\}. \tag{5.47}$$

Proposition 5.1 *The permissible region* $\widetilde{\mathbf{W}}_{Interval}$ *in (5.47) is equivalent to the permissible region* $\mathbf{W}_{Interval}$ *in (5.43).* ■

The proof of Proposition 5.1 is given in Appendix 5.6.2. Since (1) there are only finite scenarios in the linear constraints (5.44)–(5.46), and (2) Proposition 5.1 indicates the equivalence between $\widetilde{\mathbf{W}}_{\mathrm{Interval}}$ and $\mathbf{W}_{\mathrm{Interval}}$, the multi-parametric programming theory in [10] can be applied to (5.44)–(5.46) to get the permissible region of tie-line power flow exchanges when facing renewable uncertainties delineated by intervals.

5.4.2.2 Scenario-Based Decomposition Acceleration

When the number of stochastic variables $\mathbf{P}_{\mathrm{D}}$ increases, the number of the linear constraints (5.44)–(5.46) significantly increases due to the exponential increase of extreme scenarios. Consequently, the resultant computational burden to get the permissible region of tie-line power flow exchanges using multi-parametric programming also significantly increases, which may prevent its practical implementations in limited operational time. To alleviate such a potentially heavy computational burden, a scenario-based decomposition strategy that facilitates a parallel computing framework will be introduced. In the scenario-based decomposition strategy, it is assumed that generators own sufficient abilities to change generation levels among different extreme scenarios. Consequently, the linear constraints (5.44)–(5.46) can be decomposed across different extreme scenarios. For the sth scenario, its constraints after decomposition are given as

$$\mathbf{h}_1\mathbf{P}_{\mathrm{G}}^{s} + \mathbf{h}_2\mathbf{P}_{\mathrm{B}} + \mathbf{h}_3\mathbf{P}_{\mathrm{D}}^{s} = \mathbf{h}_4, \tag{5.48}$$

$$\mathbf{g}_1\mathbf{P}_{\mathrm{G}}^{s} + \mathbf{g}_2\mathbf{P}_{\mathrm{B}} + \mathbf{b} \leq \mathbf{g}_3\mathbf{P}_{\mathrm{D}}^{s}, \tag{5.49}$$

$$\text{Constraints (5.40)-(5.41)}, \tag{5.50}$$

where $\mathbf{P}_{\mathrm{G}}^{s}$ stacks generation levels in the sth scenario.

Let's denote one permissible region of tie-line power flow exchanges as $\widetilde{\mathbf{W}}_{\text{Interval},s}$ after applying the multi-parametric programming theory to the constraints (5.48)–(5.50). Then, the permissible region of tie-line power flow exchanges when facing renewable uncertainties delineated by intervals is estimated as $\widetilde{\mathbf{W}}_{\text{Interval}} = \bigcap_{s=1}^{|\mathbf{S}|} \widetilde{\mathbf{W}}_{\text{Interval},s}$ where $|\mathbf{S}|$ is the cardinality of the set $\mathbf{S}$. The acceleration from this scenario-based decomposition strategy can be observed from the two perspectives. Firstly, each $\widetilde{\mathbf{W}}_{\text{Interval},s}$ can be individually identified using multi-parametric programming, advocating a parallel computing framework for acceleration. Secondly, compared with the original constraints (5.44)–(5.46), the number of the constraints (5.48)–(5.50) is significantly reduced. The reduced number accelerates the implementation of multi-parametric programming for identifying one permissible region of tie-line power flow exchanges.

5.4.3 Case Studies

The resource aggregation methods in Sect. 5.4 will be examined in the IEEE 9-bus and IEEE 118-bus test systems. The system parameters and numerical settings are the same as those in Sect. 5.3.3.

5.4.3.1 Effectiveness of the Resource Aggregation Method in Sect. 5.4.2.1

The IEEE 9-bus test system is utilized to demonstrate the permissible region of tie-line power flow exchanges under renewable uncertainties delineated by intervals. Figure 5.5 demonstrates such a permissible region when renewable uncertainties are considered by the linear constraints with finite scenarios in (5.44)–(5.46).

The resource aggregation method in Sect. 5.4.2.1 is validated by using the points $\left(P_5^j, P_9^j\right)$ $\left(j = 1, 2, \ldots, 5 \times 10^3\right)$ that belongs to the permissible region in Fig. 5.5

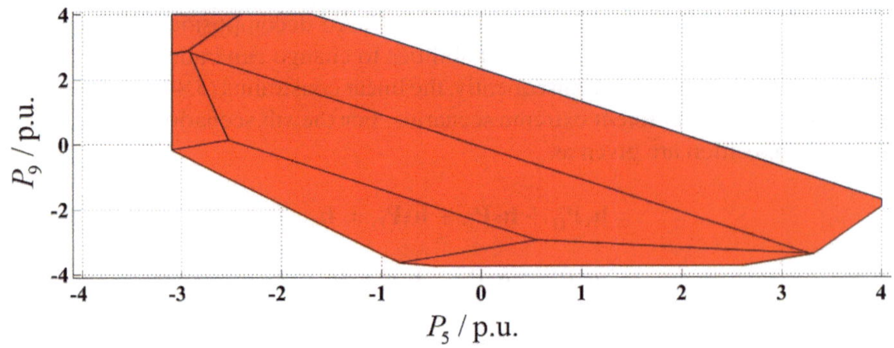

Fig. 5.5 Permissible region of tie-line power flow exchanges under renewable uncertainties delineated by intervals (IEEE 9-bus test system), reprinted from [12], open access

are generated. For a given point, the fulfillment of the constraints (5.38)–(5.41) is checked by the 5×10^3 scenarios generated from the intervals (5.42). The numerical results show that all the generated points can satisfy constraints (5.38)–(5.41). This validates the effectiveness of the resource aggregation method in Sect. 5.4.2.1 when renewable uncertainties are delineated by intervals.

It is noted that the permissible regions in Figs. 5.3 and 5.5 are similar when renewable uncertainties are described by PDFs and intervals. This could occur when two conditions hold. Firstly, when the tolerance of constraint violations is set to a small value, the impacts of the chance constraints in the resource aggregation method in Sect. 5.3.2 are similar to the impacts of the linear constraints in the resource aggregation method in Sect. 5.4.2.1. In addition, since the stochastic variables in the resource aggregation method in Sect. 5.3.2 are described by Gaussian distributions in numerical experiments, the probability where the stochastic variables located in the range $[\mu - 3\sigma, \mu + 3\sigma]$ is 99% based on the well-known 3σ-principle. Consequently, the second condition is that the range is close to the interval in the resource aggregation method in Sect. 5.4.2.1, leading to the fact that the distribution impact of stochastic variables in the resource aggregation method in Sect. 5.4.2.1 are similar to that in the resource aggregation method in Sect. 5.4.2.1.

5.4.3.2 Computational Efficiency of the Aggregation Method in Sect. 5.4.2.2

When renewable uncertainties are described by intervals, the following two methods are compared in the IEEE 118-bus test system to demonstrate the advanced computational efficiency of the resource aggregation method in Sect. 5.4.2.2:

M1: The resource aggregation method in Sect. 5.4.2.1 which employ the linear constraints (5.44)–(5.46) under joint extreme scenarios.

M2: The resource aggregation method in Sect. 5.4.2.2 which employ the linear constraints (5.48)–(5.50) under decomposed extreme scenarios.

In the numerical experiments, the computational time of the method M2 is 545 s while the method M1 cannot terminate within two hours (i.e., 7200 s). The reasons for their differences in computational time are analyzed below.

The total number of extreme scenarios is $2^4 = 16$ in the numerical experiments. In the method M1, the linear constraints (5.44)–(5.46) include 16 extreme scenarios; thus, its constraint number is rather large so that the method M1 cannot terminate within 7200 s. In contrast, the method M2 only requires the individual identification of the permissible region of tie-line power flow exchanges under each extreme scenario, and their intersection leads to the final permissible region of tie-line power flow exchanges when renewable uncertainties are delineated by intervals. A parallel computing framework can be used to accelerate the computational speed above. These issues make that the computational time of the method M2 is only 545 s, which is significantly less than that of the method M1.

5.4.3.3 Investigation of the Interval Length

When renewable uncertainties are described by intervals, a longer interval length indicates a worse prediction accuracy. The impacts of the interval length on the permissible region of tie-line power flow exchanges will be examined by computing permissible regions under the following three prediction intervals:

C1: The prediction interval is set to [**1.2** p.u., **1.25** p.u.].

C2: The prediction interval is set to [**1.0** p.u., **1.5** p.u.].

C3: The prediction interval is set to [**0** p.u., **2.5** p.u.].

The volumes of permissible regions under C1–C3 are listed in Table 5.1 . Generally, a larger volume of the permissible region of tie-line power flow exchanges indicates more opportunities for renewable accommodation through tie-line power flow exchanges among different transmission sub-systems. As shown in Table 5.1, when the prediction interval increases from [**1.2** p.u., **1.25** p.u.] to [**0** p.u., **2.5** p.u.], the volume of the permissible region decrease from 4832.723 p.u. × p.u. × p.u. to 4562.134 p.u. × p.u. × p.u. This is because a longer interval indicates more extreme scenarios which the permissible region of tie-line power flow exchanges should fulfill. In other words, more tie-line power flow exchanges should be somehow forbidden to ensure the fulfillment of transmission-level constraints in more extreme scenarios.

Table 5.1 Volumes of permissible regions under different prediction intervals, reprinted from [12], open access

Case	Volume (p.u. × p.u.× p.u.)
C1	4832.723
C2	4636.483
C3	4562.134

The results in Table 5.1 reveal that a worse prediction accuracy (i.e., a longer length of the prediction interval) reduces the volume of the permissible region of tie-line power flow exchanges. Based on the evidence above, a new perspective can be found that good prediction accuracy is beneficial for renewable accommodation through broader tie-line power flow exchanges among different transmission sub-systems.

5.5 Summary

In this chapter, the resource aggregation for low-carbon power systems under renewable uncertainties is discussed. The resource aggregation starts from the representation of renewable uncertainties. Particularly, the renewable uncertainties can be represented by either PDFs or intervals, both of which can be computed using historical observations around renewable power predictions. Since there are two representation forms of renewable uncertainties, two resource aggregation methods are accordingly introduced by taking the transmission-level resource aggregation as an example.

When renewable uncertainties are delineated by PDFs, the uncertainty consideration can be reflected in the transmission-level constraints based on chance constraints. Such chance constraints can be equivalently converted as a series of linear constraints using quantile values. This facilitates the employment of the multi-parametric programming theory to identify the permissible region of tie-line power flow exchanges under renewable uncertainties delineated by PDFs.

When renewable uncertainties are delineated by intervals, the uncertainty consideration can be reflected using a series of transmission-level constraints in extreme scenarios. This facilitates the employment of the multi-parametric programming theory to identify the permissible region of tie-line power flow exchanges under renewable uncertainties delineated by intervals. Moreover, the computational burden arising from a large number of transmission-level constraints in a large number of extreme scenarios can be alleviated by a scenario decomposition strategy. By implementing the scenario decomposition strategy, the whole permissible region in jointly extreme scenarios can be approximated as the intersection of permissible regions in each scenario.

The impacts of renewable uncertainties on the permissible region of tie-line power flow exchanges are investigated. The following two major findings are reported:

- When renewable uncertainties are delineated by PDFs, smaller confidence levels in chance constraints imply a smaller permissible region of tie-line power flow exchanges.
- When renewable uncertainties are delineated by intervals, a smaller interval length implies a larger permissible region of tie-line power flow exchanges.

5.6 Appendixes

5.6.1 Transformation Based on Cholesky Decomposition

Let's recall the symbols which will be used in this Appendix to explain the transformation based on Cholesky decomposition, including the dependent stochastic variables $\mathbf{P}_{\mathrm{D}}$ with a Pearson correlation matrix $\boldsymbol{\rho}_{\mathrm{D}} = \mathbf{G}\mathbf{G}^{\top}$, and the independent variables $\boldsymbol{\gamma}$. Also, the ith row and jth column element $\rho_{i,j}$ is defined as

$$\rho_{i,j} = \rho\left(P_{\mathrm{D},i}, P_{\mathrm{D},j}\right) = \begin{cases} \frac{\operatorname{cov}\left(P_{\mathrm{D},i}, P_{\mathrm{D},j}\right)}{\sigma_{P_{\mathrm{D},i}}\sigma_{P_{\mathrm{D},j}}}, & \text{if } i \neq j \\ 1, & \text{if } i = j \end{cases}, \tag{5.51}$$

where $P_{\mathrm{D},i}$ is the ith element in $\mathbf{P}_{\mathrm{D}}$; $\operatorname{cov}\left(P_{\mathrm{D},i}, P_{\mathrm{D},j}\right)$ is the covariance between $P_{\mathrm{D},i}$ and $P_{\mathrm{D},j}$; $\sigma_{P_{\mathrm{D},i}}$ is the standard deviation of $P_{\mathrm{D},i}$. Note that $\operatorname{cov}\left(P_{\mathrm{D},i}, P_{\mathrm{D},j}\right)$ is equal to $\operatorname{cov}\left(P_{\mathrm{D},j}, P_{\mathrm{D},i}\right)$ and thus $\rho(P_{\mathrm{D},i}, P_{\mathrm{D},j})$ is equal to $\rho(P_{\mathrm{D},j}, P_{\mathrm{D},i})$.

Suppose that there is one matrix $\mathbf{Z}$ such that $\boldsymbol{\gamma} = \mathbf{Z}\mathbf{P}_{\mathrm{D}}$. For the independent variables $\boldsymbol{\gamma}$, the corresponding Pearson correlation matrix $\boldsymbol{\rho}_{\gamma}$ is an identity matrix $\mathbf{I}$, i.e.,

$$\boldsymbol{\rho}_{\gamma} = \mathbf{I}. \tag{5.52}$$

Based on the definition of a Pearson correlation matrix, $\boldsymbol{\rho}_{\gamma}$ can be further formulated as

$$\boldsymbol{\rho}_{\gamma} = \rho\left(\boldsymbol{\gamma}, \boldsymbol{\gamma}^{\top}\right). \tag{5.53}$$

Based on $\boldsymbol{\gamma} = \mathbf{Z}\mathbf{P}_{\mathrm{D}}$ and $\boldsymbol{\rho}_{\mathrm{D}} = \boldsymbol{G}\boldsymbol{G}^{\top}$, (5.53) can be reformulated as follows:

$$\begin{aligned} \boldsymbol{\rho}_{\gamma} &= \rho\left(\gamma, \gamma^{\top}\right) \\ &= \rho\left(\mathbf{Z}\boldsymbol{P}_{\mathrm{D}}, \mathbf{P}_{\mathrm{D}}^{\top}\mathbf{Z}^{\top}\right) \\ &= \mathbf{Z}\rho\left(\mathbf{P}_{\mathrm{D}}, \mathbf{P}_{\mathrm{D}}^{\top}\right)\mathbf{Z}^{\top} \\ &= \mathbf{Z}\boldsymbol{\rho}_{\mathrm{D}}\mathbf{Z}^{\top} \\ &= (\mathbf{Z}\mathbf{G})\left(\mathbf{G}^{\top}\mathbf{Z}^{\top}\right). \end{aligned} \tag{5.54}$$

Since (5.54) should be equal to (5.52), the following equation holds:

$$(\mathbf{Z}\mathbf{G})\left(\mathbf{G}^{\top}\mathbf{Z}^{\top}\right) = \mathbf{I}. \tag{5.55}$$

It is obvious that (5.55) holds when $\mathbf{Z} = \mathbf{G}^{-1}$, i.e., $\boldsymbol{\gamma} = \mathbf{G}^{-1}\mathbf{P}_{\mathrm{D}}$. Note that $\mathbf{G}$ is a triangular decomposition of $\boldsymbol{\rho}_{\mathrm{D}}$; thus, $\mathbf{G}$ is invertible to further get $\mathbf{P}_{\mathrm{D}} = \mathbf{G}\boldsymbol{\gamma}$ from $\boldsymbol{\gamma} = \mathbf{G}^{-1}\mathbf{P}_{\mathrm{D}}$.

5.6.2 Proof of Proposition 5.1

This proof is completed by proving (1) $\mathbf{W}_{\text{Interval}} \subseteq \widetilde{\mathbf{W}}_{\text{Interval}}$ and (2) $\widetilde{\mathbf{W}}_{\text{Interval}} \subseteq \mathbf{W}_{\text{Interval}}$.

(1) Proof of $\widetilde{\mathbf{W}}_{\text{Interval}} \subseteq \mathbf{W}_{\text{Interval}}$

For the point $\mathbf{P}_{\text{B}}^{0} \in \widetilde{\mathbf{W}}_{\text{Interval}}$, there are $\mathbf{P}_{\text{G}}^{0}$ which satisfy the constraints (5.44)–(5.46), i.e.,

$$\mathbf{h}_1\mathbf{P}_{\text{G}}^{0} + \mathbf{h}_2\mathbf{P}_{\text{B}}^{0} + \mathbf{h}_3\mathbf{P}_{\text{D}}^{s} = \mathbf{h}_4,\ s \in \mathbf{S}, \tag{5.56}$$

$$\mathbf{g}_1\mathbf{P}_{\text{G}}^{0} + \mathbf{g}_2\mathbf{P}_{\text{B}}^{0} + \mathbf{b} \leq \mathbf{g}_3\mathbf{P}_{\text{D}}^{s},\ s \in \mathbf{S}, \tag{5.57}$$

$$\underline{\mathbf{P}}_{\text{G}} \leq \mathbf{P}_{\text{G}}^{0} \leq \overline{\mathbf{P}}_{\text{G}}, \tag{5.58}$$

$$\underline{\mathbf{P}}_{\text{B}} \leq \mathbf{P}_{\text{B}}^{0} \leq \overline{\mathbf{P}}_{\text{B}}. \tag{5.59}$$

Note that any point $\mathbf{P}_{\text{D}}^{0}$ in $\left[\mathbf{P}_{\text{D}}^{-}, \mathbf{P}_{\text{D}}^{+}\right]$ can be formulated as a convex combination of $\mathbf{P}_{\text{D}}^{s}$ $(s \in \mathbf{S})$ [15], i.e.,

$$\mathbf{P}_{\text{D}}^{0} = \sum_{s\in\mathbf{S}} \lambda_s \mathbf{P}_{\text{D}}^{s}, \tag{5.60}$$

$$\sum_{s\in S} \lambda_s = 1, \tag{5.61}$$

$$\lambda_s \geq 0,\ s \in S. \tag{5.62}$$

The non-negative λ_s is multiplied on both sides of the equations (5.56)–(5.57) associated with the sth extreme scenario. Adding together the results of the Eqs. (5.56)–(5.57) associated with all the extreme scenarios in $\mathbf{S}$ yields:

$$\sum_{s\in S} \lambda_s \mathbf{h}_1\mathbf{P}_{\text{G}}^{0} + \sum_{s\in S} \lambda_s \mathbf{h}_2\mathbf{P}_{\text{B}}^{0} + \sum_{s\in S} \lambda_s \mathbf{h}_3\mathbf{P}_{\text{D}}^{s} = \sum_{s\in S} \lambda_s \mathbf{h}_4, \tag{5.63}$$

$$\sum_{s\in S} \lambda_s \mathbf{g}_1\mathbf{P}_{\text{G}}^{0} + \sum_{s\in S} \lambda_s \mathbf{g}_2\mathbf{P}_{\text{B}}^{0} + \sum_{s\in S} \lambda_s \mathbf{b} \leq \sum_{s\in S} \lambda_s \mathbf{g}_3\mathbf{P}_{\text{D}}^{s}. \tag{5.64}$$

Substituting (5.60)–(5.61) into (5.63)–(5.64) yields:

$$\mathbf{h}_1\mathbf{P}_{\text{G}}^{0} + \mathbf{h}_2\mathbf{P}_{\text{B}}^{0} + \mathbf{h}_3\mathbf{P}_{\text{D}}^{0} = \mathbf{h}_4, \tag{5.65}$$

$$\mathbf{g}_1\mathbf{P}_{\text{G}}^{0} + \mathbf{g}_2\mathbf{P}_{\text{B}}^{0} + \mathbf{b} \leq g_3\mathbf{P}_{\text{D}}^{0}. \tag{5.66}$$

If there is one point $\mathbf{P}_{\text{B}}^{0} \in \widetilde{\mathbf{W}}_{\text{Interval}}$, there are $\mathbf{P}_{\text{G}}^{0}$ which can satisfy the constraints (5.40) such that the constraints (5.38)–(5.39) can be fulfilled for any $\mathbf{P}_{\text{D}}^{0}$

in $\left[\mathbf{P}_{\mathrm{D}}^{-}, \mathbf{P}_{\mathrm{D}}^{+}\right]$. Additionally, $\mathbf{P}_{\mathrm{B}}^{0} \in \widetilde{\mathbf{W}}_{\text{Interval}}$ yields the fulfillment of the constraints (5.41). Consequently, $\mathbf{W}_{\text{Interval}} \subseteq \widetilde{\mathbf{W}}_{\text{Interval}}$ holds.

(2) Proof of $\mathbf{W}_{\text{Interval}} \subseteq \widetilde{\mathbf{W}}_{\text{Interval}}$

For the point $\mathbf{P}_{\mathrm{B}}^{0} \in \mathbf{W}_{\text{Interval}}$, there are $\mathbf{P}_{\mathrm{G}}^{0}$ which satisfy the constraints (5.38)–(5.41) for any $\mathbf{P}_{\mathrm{D}}^{0}$ in $\left[\mathbf{P}_{\mathrm{D}}^{-}, \mathbf{P}_{\mathrm{D}}^{+}\right]$, i.e.,

$$\mathbf{h}_1\mathbf{P}_{\mathrm{G}}^{0} + \mathbf{h}_2\mathbf{P}_{\mathrm{B}}^{0} + \mathbf{h}_3\mathbf{P}_{\mathrm{D}}^{0} = \mathbf{h}_4, \tag{5.67}$$

$$\mathbf{g}_1\mathbf{P}_{\mathrm{G}}^{0} + \mathbf{g}_2\mathbf{P}_{\mathrm{B}}^{0} + \mathbf{b} \le \mathbf{g}_3\mathbf{P}_{\mathrm{D}}^{0}, \tag{5.68}$$

$$\underline{\mathbf{P}}_{\mathrm{G}} \le \mathbf{P}_{\mathrm{G}}^{0} \le \overline{\mathbf{P}}_{\mathrm{G}}, \tag{5.69}$$

$$\underline{\mathbf{P}}_{\mathrm{B}} \le \mathbf{P}_{\mathrm{B}}^{0} \le \overline{\mathbf{P}}_{\mathrm{B}}. \tag{5.70}$$

Since any $\mathbf{P}_{\mathrm{D}}^{0}$ in $\left[\mathbf{P}_{\mathrm{D}}^{-}, \mathbf{P}_{\mathrm{D}}^{+}\right]$ is a convex combination of $\mathbf{P}_{\mathrm{D}}^{s}$ ($s \in \mathbf{S}$), the equations (5.67)–(5.68) can be reformulated as

$$\sum_{s\in S}\lambda_s\mathbf{h}_1\mathbf{P}_{\mathrm{G}}^{0} + \sum_{s\in S}\lambda_s\mathbf{h}_2\mathbf{P}_{\mathrm{B}}^{0} + \sum_{s\in S}\lambda_s\mathbf{h}_3\mathbf{P}_{\mathrm{D}}^{s} = \sum_{s\in S}\lambda_s\mathbf{h}_4, \tag{5.71}$$

$$\sum_{s\in S}\lambda_s\mathbf{g}_1\mathbf{P}_{G}^{0} + \sum_{s\in S}\lambda_s\mathbf{g}_2\mathbf{P}_{B}^{0} + \sum_{s\in S}\lambda_s\mathbf{b} \le \sum_{s\in S}\lambda_s\mathbf{g}_3\mathbf{P}_{D}^{s}. \tag{5.72}$$

Furthermore, the constraints associated with each extreme scenario can be extracted from (5.71)–(5.72), i.e.,

$$\lambda_s\mathbf{h}_1\mathbf{P}_{\mathrm{G}}^{0} + \lambda_s\mathbf{h}_2\mathbf{P}_{\mathrm{B}}^{0} + \lambda_s\mathbf{h}_3\mathbf{P}_{\mathrm{D}}^{s} = \lambda_s\mathbf{h}_4, \forall s \in \mathbf{S}, \mathbf{S} = \mathbf{S}_1 \cup \mathbf{S}_2, \tag{5.73}$$

$$\lambda_s\mathbf{g}_1\mathbf{P}_{\mathrm{G}}^{0} + \lambda_s\mathbf{g}_2\mathbf{P}_{\mathrm{B}}^{0} + \lambda_s\mathbf{b} \le \lambda_s\mathbf{g}_3\mathbf{P}_{\mathrm{D}}^{s}, s \in \mathbf{S}, \mathbf{S} = \mathbf{S}_1 \cup \mathbf{S}_2, \tag{5.74}$$

where $\mathbf{S}_1$ is the set of the scenarios whose λ_s equals non-zero; $\mathbf{S}_2$ is the set of the scenarios whose λ_s equals zero.

For the scenarios in $\mathbf{S}_1$, both sides of the equations (5.73)–(5.74) can be divided by λ_s, i.e.,

$$\mathbf{h}_1\mathbf{P}_{\mathrm{G}}^{0} + \mathbf{h}_2\mathbf{P}_{\mathrm{B}}^{0} + \mathbf{h}_3\mathbf{P}_{\mathrm{D}}^{s} = \mathbf{h}_4, \forall s \in \mathbf{S}_1, \tag{5.75}$$

$$\mathbf{g}_1\mathbf{P}_{\mathrm{G}}^{0} + \mathbf{g}_2\mathbf{P}_{\mathrm{B}}^{0} + \mathbf{b} \le \mathbf{g}_3\mathbf{P}_{\mathrm{D}}^{s}, s \in \mathbf{S}_1. \tag{5.76}$$

For the scenarios in $\mathbf{S}_2$, the constraints (5.38)–(5.39) hold, i.e.,

$$\mathbf{h}_1\mathbf{P}_{G} + \mathbf{h}_2\mathbf{P}_{\mathrm{B}} + \mathbf{h}_3\mathbf{P}_{\mathrm{D}}^{s} = \mathbf{h}_4, s \in \mathbf{S}_2, \tag{5.77}$$

$$\mathbf{g}_1\mathbf{P}_{\mathrm{G}} + \mathbf{g}_2\mathbf{P}_{\mathrm{B}} + \mathbf{b} \le \mathbf{g}_3\mathbf{P}_{\mathrm{D}}^{s}, s \in \mathbf{S}_2. \tag{5.78}$$

Combining (5.75)–(5.78) yields:

$$\mathbf{h}_1\mathbf{P}_{\mathrm{G}}^0 + \mathbf{h}_2\mathbf{P}_{\mathrm{B}}^0 + \mathbf{h}_3\mathbf{P}_{\mathrm{D}}^s = \mathbf{h}_4, \forall s \in \boldsymbol{S}, \tag{5.79}$$

$$\mathbf{g}_1\mathbf{P}_{\mathrm{G}}^0 + \mathbf{g}_2\mathbf{P}_{\mathrm{B}}^0 + \mathbf{b} \leq \mathbf{g}_3\mathbf{P}_{\mathrm{D}}^s, s \in \mathbf{S}. \tag{5.80}$$

When $\mathbf{P}_{\mathrm{B}}^0 \in \mathbf{W}_{\mathrm{Interval}}$, there are $\mathbf{P}_{\mathrm{G}}^0$ which can satisfy the constraints (5.44)–(5.46), i.e., $\widetilde{\mathbf{W}}_{\mathrm{Interval}} \subseteq \mathbf{W}_{\mathrm{Interval}}$ holds.

References

1. The proportion of renewable installed capacities in China exceeds 40%. https://www.gov.cn/yaowen/liebiao/202409/content_6976710.htm
2. Wang, Y., Guo, C., Chen, X., et al.: Carbon peak and Carbon neutrality in China: goals, implementation path, and prospects. China Geol. **4**(4), 720–746 (2021)
3. Wang, J., Chen, L., Tan, Z., et al.: Inherent spatiotemporal uncertainty of renewable power in China. Nat. Commun. **14**, 5379 (2023)
4. Yu, J., Lin, W., Kamel, S., et al.: Equivalent model considering frequency characteristics and renewable uncertainties for probabilistic power flow. IET Gener., Transm. Distrib. **12**(22), 5939–5948 (2018)
5. Rathore, A., Kumar, A., Patel, S., et al: Modelling of wind speed data using Weibull distribution function in Dewas. In: 2022 IEEE International Power and Renewable Energy Conference (2022)
6. Hong, T., Fan, S.: Probabilistic electric Load forecasting: a tutorial review. Int. J. Forecast. **32**(3), 914–938 (2016)
7. Yin, M., Iannelli, A., Smith, R.S.: Maximum likelihood estimation in data-driven modeling and control. IEEE Trans. Autom. Control **68**(1), 317–328 (2023)
8. Hong, T., Pinson, P., Wang, Y., et al.: Energy forecasting: a review and outlook. IEEE Open Access J. Power Energy **7**, 376–388 (2020)
9. Technical requirements for dispatching side forecasting systems of wind or photovoltaic power. https://std.samr.gov.cn//gb/search/gbDetailed?id=CE1E6A1DD53258F6E05397BE0A0A68DF
10. Pistikopoulos, E., Georgiadis, M., Dua, V.: Multi-parametric Programming: Theory, Algorithms and Applications. WileyVCH, Weinheim (2007)
11. Milano, F.: Power System Modeling and Scripting. Springer, Berlin (2010)
12. Lin, W.: Characterization and fast calculation of steady-state feasible region of tie-line transmission, Ph.D. Thesis, Chongqing University (2021) (in Chinese)
13. Morales, J.M., Baringo, L., Conejo, A.J.: Probabilistic power flow with correlated wind sources. IET Gener. Transm. Distrib. **4**(5), 641–651 (2010)
14. Casedata_Uncertainties. https://figshare.com/articles/Casedata_Uncertainties/11750382
15. Boyd, S., Vandenberghe, L.: Convex Optimization. Cambridge University Press (2012)

Chapter 6
Applications in the Operations and Planning of Low-Carbon Power Systems

6.1 Introduction

Chapters 2–5 focus on the computations of permissible regions at the transmission level, and distribution level, interacted with heterogeneous energy flows, and their extension to uncertain cases. Their application examples for coordinated energy management in low-carbon power systems will be discussed in this chapter. Particularly, four application examples will be employed to elaborate on how permissible regions can contribute to coordinated energy management and what benefits can arise from the employment of permissible regions:

- The first example in Sect. 6.2 is for coordinated operations among transmission sub-systems, by leveraging permissible regions of tie-line power flow exchanges for a tie-line power flow exchange scheduling to facilitate better resource configurations at the transmission level.
- The second example in Sect. 6.3 is for coordinated operations between one transmission sub-system and its interconnected distribution sub-systems, by leveraging permissible regions of the *point of common coupling* (PCC) flow exchanges to make full use of *distributed energy resources* (DERs) at the distribution level.
- The third example in Sect. 6.4 is for *reliability-oriented distribution expansion planning* (RODP), by leveraging permissible regions for power transfers from power consumers to enhance the investment efficiency on both the distribution and consumer sides.
- The fourth example in Sect. 6.5 is for heat and electricity integration, by leveraging permissible regions to improve complementary utilization between heterogeneous energy flows.

W. Lin et al., *Coordinated Energy Management for Low-Carbon Power Systems*,
https://doi.org/10.1007/978-981-96-7533-3_6

6.2 Application Example 1. Coordinated Operations Among Transmission Sub-Systems

6.2.1 *Overview*

Different transmission sub-systems are generally operated by different operators although they are physically interconnected by tie-lines. This prevents a centralized manner for their coordinated operations by a coordination center towards renewable accommodation in low-carbon power systems due to privacy issues from different sub-system operators. Alternatively, Sect. 6.2 will employ permissible regions of tie-line power flow exchanges from Chap. 2 to facilitate such coordinated operations among different transmission sub-systems. Particularly, the resultant coordination results can be equivalent to those in a centralized manner and contain the following two stages:

Stage 1. The coordination center schedules tie-line power flow exchanges by receiving the permissible region of tie-line power flow exchanges from each transmission sub-system.

Stage 2. Given the tie-line power flow exchanges scheduled from the coordination center at Stage 2, each transmission sub-system can dispatch its individual dispatchable resources (e.g., generators) to respond to the tie-line power flow exchange scheduling.

The implementation details of the two stages will be elaborated in the coming Sect. 6.2.2.

6.2.2 *Implementation Details*

6.2.2.1 Implementations at Stage 1

Stage 1 requires the coordination center to schedule tie-line power flow exchanges after receiving permissible regions of all the transmission sub-systems. This task is first resolved by formulating the *economic dispatch* (ED) problem (2.9) and (2.10) of one transmission sub-system in the domain of tie-line power flow exchanges. Particularly, this can be completed by using permissible regions of tie-line power flow exchanges from Chap. 2, i.e., the ED problem (2.9) and (2.10) of one transmission sub-system can be reformulated as

$$\min_{\mathbf{w},k_i,\forall i} \sum_{i=1}^{\mathrm{N}} k_i z_i^*(\mathbf{w}), \tag{6.1}$$

$$\text{s.t.} \quad k_i \mathbf{G}_{\mathrm{w},i} \mathbf{w} \le k_i \mathbf{F}_{\mathrm{w},i},\ i = \{1, 2, \ldots, \mathrm{N}\}, \tag{6.2}$$

$$z_i^*(\mathbf{w}) = \mathbf{G}_{\mathrm{z},i} \mathbf{w} + \mathbf{F}_{\mathrm{z},i},\ i = \{1, 2, \ldots, \mathrm{N}\}, \tag{6.3}$$

$$\sum_{i=1}^{\mathrm{N}} k_i = 1, \tag{6.4}$$

$$k_i = \{0, 1\}, i = \{1, 2, \ldots, \mathrm{N}\}, \tag{6.5}$$

where $\mathbf{G}_{\mathrm{w},i}$ and $\mathbf{F}_{\mathrm{w},i}$ are the parameters from Chap. 2 through (2.23) to determine the ith critical region; $\mathbf{G}_{\mathrm{z},i}$ and $\mathbf{F}_{\mathrm{z},i}$ are the parameters from Chap. 2 through (2.25) to reflect the minimum operational cost for the response to tie-line power flow exchanges $\mathbf{w}$; the binary variable k_i is employed to justify whether tie-line power flow exchanges $\mathbf{w}$ fall into the ith critical region. Particularly, the constraint (6.4) ensures that tie-line power flow exchanges fall into only one unique critical region.

Although the optimization problem (6.1)–(6.5) can reformulate the ED problem (2.9)–(2.10) in the domain of tie-line power flow exchanges $\mathbf{w}$, the optimization problem (6.1)–(6.5) is a *mixed-integer nonlinear programming* (MINLP) problem and thus difficult to ensure its convergence and optimality in limited operational time. Consequently, the coordination center may face computational challenges in tie-line power flow exchange scheduling through the optimization problem (6.1)–(6.5) from each transmission sub-system. To resolve this issue, the optimization problem (6.1)–(6.5) is suggested to be equivalently formulated as the following *linear programming* (LP) problem:

$$\min_{y,\mathbf{w}} y, \tag{6.6}$$

$$\text{s.t} \quad y \geq z_i^*(\mathbf{w}), i = \{1, 2, \ldots, \mathrm{N}\}, \tag{6.7}$$

$$z_i^*(\mathbf{w}) = \mathbf{G}_{\mathrm{z},i}\mathbf{w} + \mathbf{F}_{\mathrm{z},i}, i = \{1, 2, \ldots, \mathrm{N}\}, \tag{6.8}$$

$$\mathbf{G}_{\mathrm{w}}\mathbf{w} \leq \mathbf{F}_{\mathrm{w}}, \tag{6.9}$$

where $\mathbf{G}_{\mathrm{w}}$ and $\mathbf{F}_{\mathrm{w}}$ are the parameters from Chap. 2 through (2.24) to represent the permissible region of tie-line power flow exchanges which is the union of all the critical regions; the continuous variable y is introduced to represent the operational cost of the transmission sub-system.

The equivalence between the MINLP problem (6.1)–(6.5) and the LP problem (6.6)–(6.9) is proven in Appendix 6.7.1. After collecting the LP problem (6.6)–(6.9) from each transmission sub-system, the coordination center can formulate the following LP problem for tie-line power flow scheduling by solving which the operational costs among all the transmission sub-systems are minimized:

$$\min_{\{y_r,\mathbf{w}_r\},\forall r} \sum_{r=1}^{\mathrm{N_R}} y_r, \tag{6.10}$$

$$\text{s.t.} \quad y_r \geq z_{r,i}^*(\mathbf{w}_r), r = \{1, 2, \ldots, \mathrm{N_R}\}, i = \{1, 2, \ldots, \mathrm{N}_r\} \tag{6.11}$$

$$z^*_{r,i}\left(\mathbf{w}_r\right) = \mathbf{G}_{zr,i}\mathbf{w}_r + \mathbf{F}_{z,r,i}, r = \{1, 2, \ldots, \mathrm{N_R}\}, i = \{1, 2, \ldots, \mathrm{N}_r\} \tag{6.12}$$

$$\mathbf{G}_{\mathrm{w},r}\mathbf{w}_r \leq \mathbf{F}_{\mathrm{w},r}, r = \{1, 2, \ldots, \mathrm{N_R}\} \tag{6.13}$$

$$\mathbf{A}_r\mathbf{w}_r + \mathbf{A}_k\mathbf{w}_k = \mathbf{0}, r, k = \{1, 2, \ldots, \mathrm{N_R}\}, r \neq k, \tag{6.14}$$

where the subscript r indicates the rth transmission sub-system.

The objective function (6.10) seeks to minimize the total operational cost of all the transmission sub-systems. The specific meaning of the constraints (6.11) can be reviewed in Fig. 6.12. The constraints (6.12) describe the piece-wise mapping relationship between the minimum operational cost of the rth transmission sub-system and tie-line power flow exchanges. The constraints (6.13) delineate the permissible region of tie-line power flow exchanges. The constraints (6.14) define the coupling constraints among different transmission sub-systems, such as the power flow consistency on the same tie-lines.

6.2.2.2 Implementations at Stage 2

After the coordination center solves the LP problem (6.6)–(6.9), tie-line power flow exchanges of each transmission sub-system can be determined. Given such a tie-line power flow exchange scheduling, each transmission sub-system can solve the following ED problem to dispatch its individual dispatchable resources for the response to the tie-line power flow exchange scheduling:

$$\min_{P_g, \forall g} z_{\mathrm{QP}} = \sum_{g=1}^{\mathrm{N_G}} \left(a_g P_g^2 + b_g P_g + c_g\right), \tag{6.15}$$

$$\text{s.t.} \quad \mathbf{MP}_{\mathrm{G}} \leq \mathbf{N}, \tag{6.16}$$

where a_g and b_g are the cost coefficients of the quadratic and linear terms of unit g, respectively; c_g is the cost coefficient of the constant term of unit g; $\mathbf{MP}_{\mathrm{G}} \leq \mathbf{N}$ delineates the constraints from (6.6)–(6.9) given the tie-line power flow scheduling.

Due to the non-linearity of the quadratic cost function in (6.15), its quadratic term $a_g P_g^2$ is usually displaced with a piece-wise linear function in current power industries [1], as shown in Fig. 6.1. In Fig. 6.1, q_g^{L} is the piece-wise linear approximation to the quadratic term $a_g P_g^2$ where k_g^i and b_g^i are the slope and intercept of the ith linear function ($i = 1, 2, \ldots, \mathrm{N_S}$, $\mathrm{N_S}$ is the number of piece-wise linearizations).

Based on Fig. 6.1, the *quadratic programming* (QP) problem (6.15)–(6.16) is approximated as

$$\min_{\{P_g, q_k^{\mathrm{L}}\}, \forall g} z_{\mathrm{LP}} = \sum_{g=1}^{\mathrm{N_G}} \left(q_g^{\mathrm{L}} + b_g P_g + c_g\right), \tag{6.17}$$

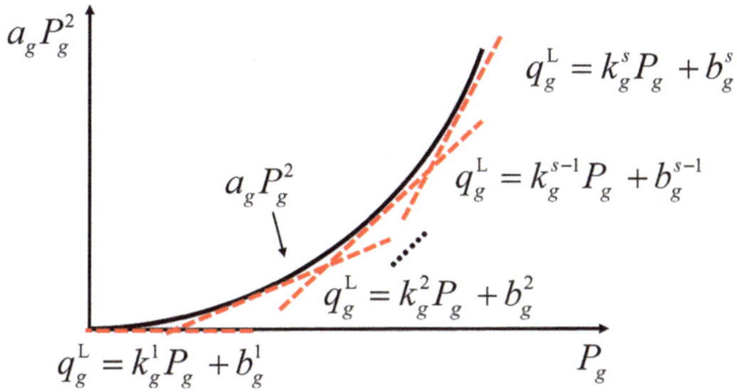

Fig. 6.1 Piece-wise linearization of a quadratic cost function, reprinted from [2], with permission from IEEE

$$\text{s.t.}\quad q_g^{\mathrm{L}} \geq k_g^i P_g + b_g^i, i = \{1, 2, \ldots, \mathrm{N_S}\}, \tag{6.18}$$

$$\mathbf{MP}_{\mathrm{G}} \leq \mathbf{N}. \tag{6.19}$$

It is obvious that the optimal solutions of the LP problem (6.17)–(6.19) are the same as the optimal solutions of the QP problem (6.15) and (6.16). However, $\mathrm{N_S}$ is usually set to a certain number (e.g., the number is 50 in MISO). Consequently, the optimal solutions of the LP problem (6.17)–(6.19) inevitably deviate from the optimal solutions of the QP problem (6.15) and (6.16). This necessitates a correct strategy by which the optimal solutions of the QP problem (6.15) and (6.16) can be recovered from the optimal solutions of the LP problem (6.15) and (6.16). To construct such a correct strategy, let's first analyze the following *Karush–Kuhn–Tucker* (KKT) conditions of the QP problem (6.15) and (6.16):

$$\frac{\partial z_{\mathrm{QP}}}{\partial \mathbf{P}_{\mathrm{G}}} + \mathbf{M}^{\top}\boldsymbol{\lambda} = \mathbf{0}, \tag{6.20}$$

$$\boldsymbol{\lambda} \otimes (\mathbf{MP}_{\mathrm{G}} - \mathbf{N}) = \mathbf{0}, \tag{6.21}$$

$$\boldsymbol{\lambda} \geq \mathbf{0}, \tag{6.22}$$

where $\boldsymbol{\lambda}$ are the dual variables of the constraints (6.16).

When active and inactive constraints are known, the KKT conditions (6.20)–(6.22) can be reformulated as

$$\frac{\partial z_{\mathrm{QP}}}{\partial \mathbf{P}_{\mathrm{G}}} + \mathbf{M}_{\mathrm{A}}^{\top}\boldsymbol{\lambda}_{\mathrm{A}} + \mathbf{M}_{\mathrm{I}}^{\top}\boldsymbol{\lambda}_{\mathrm{I}} = \mathbf{0}, \tag{6.23}$$

$$\mathbf{M}_{\mathrm{A}}\mathbf{P}_{\mathrm{G}} = \mathbf{N}_{\mathrm{A}}, \tag{6.24}$$

$$\boldsymbol{\lambda}_{\mathrm{K}} = \mathbf{0}, \tag{6.25}$$

$$\mathbf{M}_{\mathrm{I}}\mathbf{P}_{\mathrm{G}} < \mathbf{N}_{\mathrm{I}}, \tag{6.26}$$

$$\boldsymbol{\lambda}_{\mathrm{A}} \geq \mathbf{0}, \tag{6.27}$$

where the subscripts A and I are the indexes for the active and inactive constraint sets, respectively.

When active and inactive constraints are given, the linear equations (6.23)–(6.25) can be solved by regarding $(\mathbf{P}_{\mathrm{G}}, \boldsymbol{\lambda}_{\mathrm{A}})$ as variables such that the solutions $\left(\mathbf{P}_{\mathrm{G}}^{\mathrm{S}}, \boldsymbol{\lambda}_{\mathrm{A}}^{\mathrm{S}}\right)$ are obtained. When the given active and inactive constraints are the same as those of the QP problem (6.15)–(6.16) at optimum, all the KKT conditions (6.23)–(6.27) can be obtained, i.e., the optimal solutions of the QP problem (6.15)–(6.16) can be known. Consequently, the LP problem (6.17)–(6.19) can be employed to predict the active and inactive constraints of the QP problem (6.15)–(6.16). If the prediction is correct, the optimal solutions of the QP problem (6.15)–(6.16) can be recovered by solving the linear equations (6.23)–(6.25); otherwise, the solutions from the linear equations (6.23)–(6.25) will violate the inequalities (6.26)–(6.27) and thus the number of piece-wise linearizations need to increase. In summary, the following three-step correction strategy can be utilized to recover the optimal solutions of the QP problem (6.15)–(6.16) from the solutions of the LP problem (6.17)–(6.19):

Step 1: Solve the LP problem (6.17)–(6.19) to get (1) its optimal solutions $\mathbf{P}_{\mathrm{G}}^{\mathrm{LP}}$, (2) active constraint set A, and 3) inactive constraint set I.

Step 2: Solve the linear equations (6.23)–(6.25) to get its solutions $\left(\mathbf{P}_{\mathrm{G}}^{\mathrm{S}}, \boldsymbol{\lambda}_{\mathrm{J}}^{\mathrm{S}}\right)$.

Step 3: If $\left(\mathbf{P}_{\mathrm{G}}^{\mathrm{S}}, \boldsymbol{\lambda}_{\mathrm{A}}^{\mathrm{S}}\right)$ fulfills the inequalities (6.26)–(6.27), the optimal solutions of the QP problem (6.15)–(6.16) are $\mathbf{P}_{\mathrm{G}}^{\mathrm{S}}$; otherwise, increase the number of piece-wise linearizations and go back to Step 1.

The mentioned-above correction strategy also provides a new perspective on setting the number of piece-wise linearizations in the LP problem (6.17)–(6.19). It is commonly believed that a larger number of piece-wise linearizations indicates a higher linearization accuracy for the quadratic term in (6.15). This necessitates a large number of piece-wise linearizations, leading to a large number of constraints (6.18) which may make a heavy computational burden. However, based on the correction strategy above, it can be seen that when the number of piece-wise linearizations increases to a certain number, the LP problem (6.17)–(6.19) can identify the active and inactive constraints same as those of the QP problem (6.15)–(6.16) at optimum. This allows the computation of the optimal solutions of the QP problem (6.15)–(6.16) using the correction strategy. In other words, further increasing the number of piece-wise linearizations only adds a computational burden without improving solution accuracy, since the optimal solutions of the QP problem (6.15)–(6.16) can be obtained already.

Table 6.1 Comparison of tie-line power flow exchange scheduling under different methods, reprinted from [3], open access

Method	P_5(p.u.)	P_9(p.u.)	P_{60}(p.u.)
The application method in Sect. 6.2.2.1	0.399	1.060	0.146
Centralized manner	0.399	1.060	0.146

6.2.3 *Case Studies*

The application method in Sect. 6.2.2.1 is demonstrated in a two-region transmission sub-system that comes from a 661-bus utility system in China, as shown in Fig. 6.2. Its tie-line power flow scheduling is compared with that in a centralized manner in Table 6.1. It is found that the tie-line power flow scheduling of the application method in Sect. 6.2.2.1 is the same as that in a centralized manner. Different from the centralized manner where all the parameters to formulate the ED problem of each transmission sub-system should be submitted to the coordination center, the application method in Sect. 6.2.2.1 only requires the submission of the information about permissible region of tie-line power flow exchanges. Consequently, the application method in Sect. 6.2.2.1 is more suitable in practice, while the accuracy of the tie-line power flow scheduling is not affected.

The correction strategy in Sect. 6.2.2.2 is further applied to the two transmission sub-systems in Fig. 6.2 when the tie-line power flow scheduling is given. The correction results are listed in Table 6.2. It is found that (1) the operational cost after the correction strategy in Sect. 6.2.2.2 is the same as those of the QP problem (6.15)–(6.16), and (2) the operational cost of the LP problem (6.17)–(6.19) inevitably deviates from that of the QP problem (6.15)–(6.16). To more clearly investigate (1) the deviation between the QP problem (6.15)–(6.16) and the LP problem (6.17)–(6.19), and (2) the situations after applying the correction strategy, let's further conduct them in the IEEE 30-bus test system.

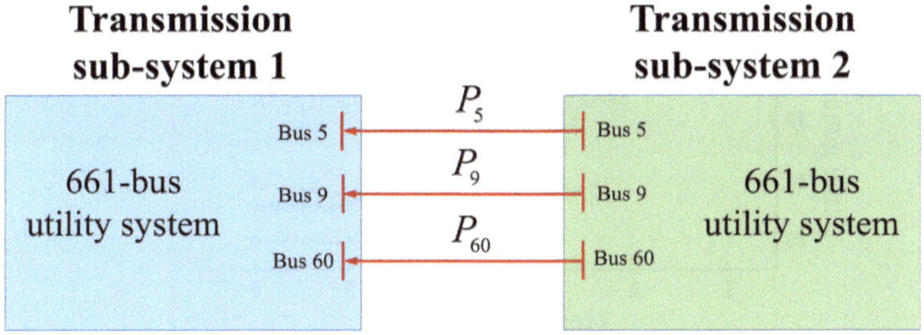

Fig. 6.2 A two-region transmission sub-system, reprinted from [3], open access

Table 6.2 Operational costs of individual transmission sub-systems, reprinted from [3], open access

Transmission sub-system	Operational cost ($/MW)		
	The QP problem (6.15)–(6.16)	The LP problem (6.17)–(6.19)	After correction
Sub-system 1	7.9493×10^4	7.9358×10^4	7.9493×10^4
Sub-system 2	8.4747×10^4	8.4643×10^4	8.4747×10^4

Figure 6.3 employs generation levels of all the units in the IEEE 30-bus test system to demonstrate the deviation between the QP problem (6.15)–(6.16) and the LP problem (6.17)–(6.19). From Fig. 6.3, it can be seen that as the number of piece-wise linearizations increases, the optimal solutions of the LP problem (6.17)–(6.19) and the QP problem (6.15)–(6.16) become closer. Particularly, there is no branch congestion and no units reach their either minimum or maximum bounds in the QP problem (6.15)–(6.16). However, for the one-segment and two-segment linearizations, there is branch congestion at the optimum of the LP problem (6.17)–(6.19), which leads to non-positive dual multipliers in the correction strategy in Sect. 6.2.2.2. Consequently, it is necessary to increase the number of piece-wise linearizations in the correction strategy in Sect. 6.2.2.2. When the number of piece-wise linearizations is set to three, the LP problem (6.17)–(6.19) still does not provide the optimal solutions of the QP problem (6.15)–(6.16). However, it allows the correction strategy to recover the optimal solutions of the QP problem (6.15)–(6.16). This is because the three-segment linearization can provide the active and inactive constraints which are the same as those of the QP problem (6.15)–(6.16) at optimum. If the number of piece-

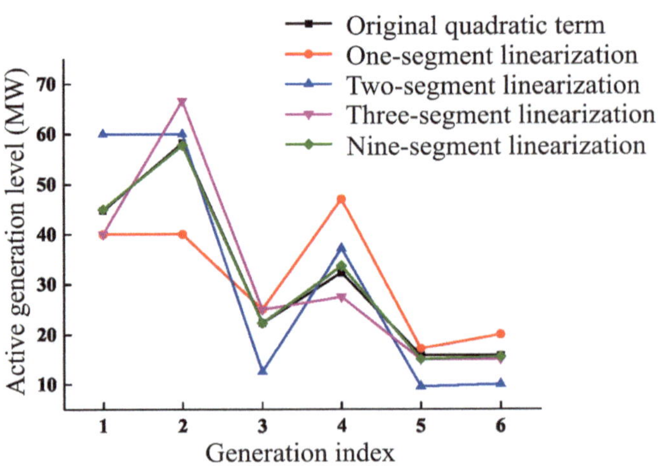

Fig. 6.3 Active generation levels with different numbers of piece-wise linearizations in the IEEE 30-bus test system, reprinted from [2], with permission from IEEE

wise linearizations continues to increase in the LP problem (6.17)–(6.19), the optimal solutions of the LP problem (6.17)–(6.19) and the QP problem (6.15)–(6.16) become increasingly closer, while the computational burden of the LP problem (6.17)–(6.19) also increases. Furthermore, the increasing number of piece-wise linearizations no longer improves computational accuracy, since the correction strategy in Sect. 6.2.2.2 can already provide the optimal solutions of the QP problem (6.15)–(6.16) based on the three-segment linearizations in the LP problem (6.17)–(6.19). in Sect. 6.2.2.2

6.3 Application Example 2. Coordinated Operations Between One Transmission Sub-System and Distribution Sub-Systems

6.3.1 Overview

Although one transmission sub-system is physically connected to distribution sub-systems at PCC, one transmission sub-system is generally operated by one transmission operator while different distribution sub-systems are operated by different distribution operators. This leads to a similar issue to the application example in Sect. 6.2 that a centralized manner cannot be applied to the coordinated operations between one transmission sub-system and its interconnected distribution sub-systems. Alternatively, Sect. 6.3.2 will employ permissible regions of PCC power flow exchanges from Chap. 3 to facilitate such coordinated operations between one transmission sub-system and its distribution sub-systems. This can be concluded that each distribution operator computes its permissible region of PCC power flow exchanges and then submits the result to the transmission operator for the transmission-level participation, as will be demonstrated in Sect. 6.3.2.

6.3.2 Implementation Details

For convenience of discussions, let's consider the coordination of one distribution sub-system with the transmission sub-system, while its generalization to multiple distribution sub-systems is straightforward. Chapter 3 shows that several inner sub-regions will be computed and their union will serve as an estimation of the permissible region of PCC power flow exchanges.

Assume $\mathrm{N_J}$ sub-regions are found from Chap. 3, and the jth sub-region $\mathbf{W}_j$ is denoted as $\mathbf{A}_j\mathbf{w} \leqslant \mathbf{B}_j$, where $\mathbf{u}_{\mathrm{PCC}}$ is PCC power flow exchanges, and the constant matrix $\mathbf{A}_j$ and the constant vector $\mathbf{B}_j$ are obtained from Chap. 3. Additionally, a distribution sub-system can be allowed to bid its PCC power transfers in a transmission sub-system using a piece-wise linear convex function $z_j(\mathbf{w})$ [4].

The incorporation of the sub-regions and their bidding cost functions into transmission-level operations can be achieved by introducing binary variables to determine a sub-region in which PCC power flow exchanges are located. Based on this idea, the following optimization problem can be formulated for the coordinated operations between one transmission sub-system and distribution sub-systems based on permissible regions of PCC power flow exchanges:

$$\min_{k_j,\mathbf{w},\mathbf{s}} C(\mathbf{s}) + \sum_{j=1}^{\mathrm{N_J}} \left(k_j z_j(\mathbf{w})\right) \tag{6.28}$$

$$\text{s.t.}\,\mathbf{g}_{\mathrm{e}}(\mathbf{s}, \mathbf{w}) = \mathbf{0}, \tag{6.29}$$

$$\mathbf{g}_{\mathrm{ie}}(\mathbf{s}, \mathbf{w}) \leq \mathbf{0}, \tag{6.30}$$

$$k_j \mathbf{A}_j \mathbf{w} \leq k_j \mathbf{B}_j,\ j = \{1, 2, \ldots, \mathrm{N_J}\}, \tag{6.31}$$

$$\sum_{j=1}^{\mathrm{N_J}} k_j = 1, \tag{6.32}$$

$$k_j = \{0, 1\},\ j = \{1, 2, \ldots, \mathrm{N_J}\}, \tag{6.33}$$

where $\mathbf{s}$ stacks the dispatchable resources of the transmission operator; $C(\mathbf{s})$ is a convex cost function of the dispatchable resources $\mathbf{s}$; $\mathbf{g}_{\mathrm{e}}$ and $\mathbf{g}_{\mathrm{ie}}$ represent the equality and inequality requirements in the transmission sub-system, respectively.

The objective function (6.28) is to minimize the total operational cost on the transmission and distribution sides. The constraints (6.29) and (6.30) are the compact formulations of transmission-level equalities (e.g., power balance) and inequalities (e.g., flow limits), respectively. The constraints (6.31) represents the jth sub-region from Chap. 3. The constraints (6.32)–(6.33) guarantee the unique selection of a sub-region. In power industries, the DC power flow model is usually employed by the transmission operator to formulate $\mathbf{g}_{\mathrm{e}}$ and $\mathbf{g}_{\mathrm{ie}}$ [5]. In this context, the optimization problem (6.28)–(6.33) is an MINLP which is difficult to solve. Consequently, a big-M formulation is employed below to linearize the MINLP problem (6.28)–(6.33).

$$\min_{y_j,\mathbf{w},\mathbf{s}} C(\mathbf{s}) + \sum_{j=1}^{n_J} y_j, \tag{6.34}$$

$$\text{s.t. Constraints (6.29)–(6.30) and (6.32)–(6.33),} \tag{6.35}$$

$$\mathbf{A}_j \mathbf{w} \leq \mathbf{B}_j + \left(1 - k_j\right) \mathrm{M},\ j = \{1, 2, \ldots, \mathrm{N_J}\}, \tag{6.36}$$

$$z_j(\mathbf{w}) \leq y_j + \left(1 - k_j\right) \mathrm{M},\ j = \{1, 2, \ldots, \mathrm{N_J}\}, \tag{6.37}$$

$$0 \leq y_j, j = \{1, 2, \ldots, \mathrm{N_J}\}, \tag{6.38}$$

where y_j is a real-valued ancillary variable; M is a real-valued number that is positive and sufficiently large.

The optimization problem (6.34)–(6.38) is a *mixed-integer linear programming* (MILP) problem instead of an MINLP problem. It is not hard to justify that the MILP problem (6.34)–(6.38) is equivalent to the MINLP problem (6.28)–(6.33), i.e., the coordination results from the MINLP problem (6.28)–(6.33) are the same as those from the MILP problem (6.34)–(6.38). Particularly, $\sum_{j=1}^{\mathrm{N_J}} y_j$ represents the total bidding cost of all the distribution sub-systems.

6.3.3 Case Studies

The application method in Sect. 6.3.2 to coordinate distribution sub-systems and the transmission sub-system will be verified using ten mirrored IEEE 136-bus test systems and a 661-bus utility system in China [6].

Two problems are first solved for the coordination between ten distribution sub-systems and the transmission sub-system: one is the MINLP problem (6.28)–(6.33), and the other is the MILP problem (6.34)–(6.38). In this testing, 1438 continuous variables, 700 binary variables, and 23692 constraints are necessary to formulate the MINLP problem (6.28)–(6.33). The scale of the MILP problem (6.34)–(6.38) is similar to that of the MINLP problem (6.28)–(6.33). It is found that solving the MILP problem (6.34)–(6.38) requires more than half an hour, which exceeds the computational window allowed in operations of certain power systems (e.g., 15 min in Guangdong Power Grid [7]). In contrast, the MILP problem (6.34)–(6.38) can solved within 539.6 s.

Based on the MILP problem (6.34)–(6.38), the impacts of permissible regions of PCC power flow exchanges on the coordination results are compared. The permissible region can be either obtained from Chap. 3 or the conservative method in [8] which employs a specific rectangle shape to ensure its inner estimation to the permissible region of PCC power flow exchanges. Their resultant PCC power flow exchange scheduling is listed in Table 6.3. Note that negative power indicates that the PCC power flow is exchanged from the distribution sub-system to the transmission sub-system. Generally, the bidding cost to dispatch DERs in distribution sub-systems will be cheaper than that to dispatch transmission-level resources (e.g., conventional units), as set in the numerical experiments. Intuitively, the transmission sub-system would like to make full use of distribution sub-systems to reduce the operational cost. Consequently, more power flow exchanges are expected from the distribution sub-systems to the transmission sub-system. As seen in Table 6.3, the total active power flow exchanges based on the permissible region from Chap. 3 is –206.84 MW, while that from [8] is –182.83 MW. This indicates that the permissible region from Chap. 3 can facilitate more power flow exchanges from distribution sub-systems, while the permissible region from the method in [8] somehow restricts the PCC

Table 6.3 Power transfer scheduling of ten distribution sub-systems when permissible regions of PCC power flow exchanges are delineated under different methods, reprinted from [9], with permission from Elsevier

Distribution sub-system	Coordination results based on the method from Chap. 3		Coordination results based on the method from [8]	
	P_{PCC} (MW)	Q_{PCC} (MVar)	P_{PCC} (MW)	Q_{PCC} (MVar)
1	−20.48	−12.78	−21.79	−10.56
2	−37.59	−13.97	−37.58	−13.97
3	42.80	36.56	42.80	36.46
4	0.85	−38.35	0.05	−37.27
5	0.16	37.98	0.88	40.00
6	−4.54	−36.07	−3.56	−38.80
7	−25.67	−30.96	−20.45	−32.79
8	−23.43	10.98	−14.93	22.24
9	−13.43	−35.97	−11.31	−35.02
10	−39.91	6.91	−31.34	13.98
Total	−206.84	−12.78	−182.83	−55.73

power flow exchanges by employing specific box-shaped sets as an estimation to the true permissible region PCC power flow exchanges. Note that such specific shapes are not required in Chap. 3, and thus its resultant estimation provides a more flexible manner to delineate the permissible region. Such flexibility improvements can reduce the total operational cost from \$$1.0873 \times 10^7$ to \$$1.0852 \times 10^7$, i.e., \$$2.1 \times 10^4$ can be saved.

6.4 Application Example 3. Reliability-Oriented Distribution Expansion Planning

6.4.1 Overview

Since distribution sub-systems are connected to power consumers, the reliability of distribution sub-systems directly affects power consumers. This necessitates RODP to improve the power supply reliability for power consumers. Conventionally, RODP is implemented by new lines and substations invested by distribution operators [10]. In the context of deregulation environments, *energy storage* (ES) invested by power consumers are also allowed in RODP due to several resultant benefits (e.g., smoothing load curves by shaving peaks and filling valleys) [11]. Consequently, The reliability of one distribution sub-system can be enhanced through both line/substation investments and power transfers from ES in power consumers, as shown in Fig. 6.4.

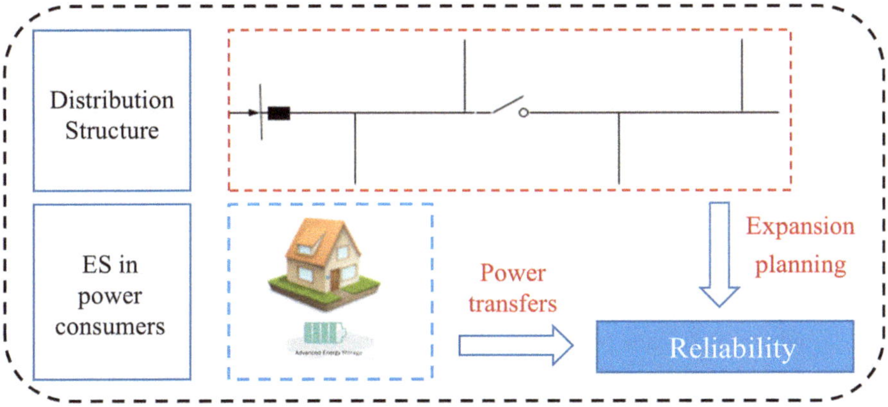

Fig. 6.4 Illustration of impact factors on reliability, reprinted from [12], open access

However, privacy issues may exist due to different stakeholders such that ES invested by power consumers can not be directly incorporated into RODP which is implemented by the distribution operator to enhance reliability. This drives the application of the resource aggregation for RODP considering ES in power consumers. The key lies in identifying the permissible region of power transfers from power consumers and then incorporating the result into RODP, as will be elaborated in Sect. 6.4.2 sequentially.

6.4.2 *Implementation Details*

6.4.2.1 Permissible Region of Power Transfers from Power Consumers

The permissible region of power transfers from one power consumer starts from its following ES planning problem:

(1) **Objective function**:

$$\min_{P_{\max}, E_{\max}, P_{d,t}, P_{c,t}, P_{\text{grid},}, S_t, \forall t} C = \sum_{i=1}^{Y} (C_P P_{\max} + C_E E_{\max}) \left(\frac{1+R_1}{1+R_2}\right)^i + \sum_{i=1}^{Y} C_M P_{\max} \left(\frac{1+R_1}{1+R_2}\right)^i, \tag{6.39}$$

where C is the total investment cost whose the first item is the investment cost and the second item is the operation and maintenance cost; C_P and C_E are the unit cost of charging/discharging power and the unit cost of capacities, respectively; C_M is the annual operation and maintenance cost of unit charging/discharging power; R_1 and R_2 are the inflation and discount rates, respectively; P_{max} is the rated charging/discharging power of installed ES; E_{max} is the rated charging/discharging capacity of installed ES; Y is the planning horizon; $t = \{1, 2, \ldots, N_T\}$ is the time slot.

(2) **Planning constraints**:

(1) One power consumer should keep its power balance, i.e.,

$$P_{load,t} - P_{grid,t} = P_{d,t} - P_{c,t}, t = \{1, 2, \ldots, N_T\}, \tag{6.40}$$

where $P_{grid,t}$ are power transfers with the distribution sub-system at time period t; $P_{load,t}$ are power demands at time period t; $P_{d,t}$ and $P_{c,t}$ are discharging and charging power, respectively.

(2) Charging and discharging power cannot exceed their rated values, i.e.,

$$\begin{cases} 0 \le P_{d,t} \le P_{max} \\ 0 \le P_{c,t} \le P_{max} \end{cases}, t = \{1, 2, \ldots, N_T\}. \tag{6.41}$$

(3) The total discharging power should not exceed the rated capacity, and the rated capacity is proportional to the rated power, i.e.,

$$\begin{cases} \sum_{t \in T} P_{d,t} \Delta t \le E_{max} \\ E_{max} = \beta P_{max} \end{cases}, t = \{1, 2, \ldots, N_T\}, \tag{6.42}$$

where Δt is the time interval between two consecutive periods; β is the energy multiple coefficient.

(4) Charging and discharging power affect the *state of charge* (SoC), i.e.,

$$S_t = S_{t-1} + \left(\eta_d P_{d,t} - \eta_c P_{c,t}\right) \Delta t, t = \{1, 2, \ldots, N_T\}, \tag{6.43}$$

where S_t is SoC at time period t; η_d and η_c are discharging and charging efficiency coefficients, respectively.

(5) SoC should not exceed its upper and lower bounds, i.e.,

$$S_{min} \le S_t \le S_{max}, t = \{1, 2, \ldots, N_T\}, \tag{6.44}$$

where S_{min} and S_{max} are lower and upper bounds, respectively.

(6) The following complementarity constraints should hold to ensure that ES cannot simultaneously charge and discharge, i.e.,

$$P_{d,t} \times P_{c,t} = 0, t = \{1, 2, \ldots, N_T\}. \tag{6.45}$$

The permissible region of power transfers from one power consumer delineates the feasible set of $P_{grid}(t = \{1, 2, \ldots, N_T\})$ with which the constraints (6.40)–(6.45) can be fulfilled. By submitting the permissible regions of power transfers from power consumers to the distribution sub-system, RODP can be completed by considering ES in power consumers when privacy issues can be avoided in the absence of detailed parameters in (6.40)–(6.45). In addition, the total cost (6.39) should also be formulated as a function of $P_{\text{grid}}(t \in \mathbf{T})$ so that RODP considering ES in power consumers can minimize the total cost on both the distribution and power consumer sides. These necessitate the multi-parametric programming theory in [13] throughout this book. If the multi-parametric programming theory can be applied to the ES planning problem (6.39)–(6.45) by regarding $P_{\text{grid},t}(t = \{1, 2, \ldots, \text{N}_\text{T}\})$ as parameters, the permissible region of power transfers from one power consumers and the function between the total cost (6.40) and $P_{\text{grid},t}(t = \{1, 2, \ldots, \text{N}_\text{T}\})$ can be obtained, similarly done in Sect. 6.2. However, the complementarity constraints (6.45) are non-linear, which prevents the employment of the multi-parametric programming theory. This motivates the relaxation method in Appendix 6.7.2 to equivalently reformulate the ES planning problem as the following LP problem:

$$\min_{\mathbf{x},\mathbf{P}_{\text{grid}},\mathbf{P}_\text{d},\mathbf{P}_\text{c}} z = \mathbf{A}_1^\top \mathbf{x} + \mathbf{A}_2^\top (\mathbf{P}_\text{d} + \mathbf{P}_\text{c}), \tag{6.46}$$

$$\text{s.t. } \mathbf{B}_1^\top \mathbf{x} + \mathbf{B}_2^\top \mathbf{P}_{\text{grid}} + \mathbf{B}_3^\top \mathbf{P}_\text{d} + \mathbf{B}_4^\top \mathbf{P}_\text{c} \le \mathbf{B}_5^\top, \tag{6.47}$$

where $\mathbf{x} = \left[P_{\max}, E_{\max}, S_1, S_2, \cdots, S_{\text{N}_\text{T}}\right]^\top$; $\mathbf{P}_{\text{grid}} = \left[P_{\text{grid},1,}, P_{\text{grid},2}, \cdots, P_{\text{grid},\text{N}_\text{T}}\right]^\top$; $\mathbf{P}_\text{c} = \left[P_{\text{c},1}, P_{\text{c},2}, \ldots, P_{\text{c},\text{N}_\text{T}}\right]^\top$; $\mathbf{P}_\text{d} = \left[P_{\text{d},1}, P_{\text{d},2}, \ldots, P_{\text{d},\text{N}_\text{T}}\right]^\top$; the vector $\mathbf{A}_1$ describes the total cost (6.39) as $\mathbf{A}_1^\top \mathbf{x}$; the vector $\mathbf{A}_2$ is filled with sufficiently small positive non-negative numbers; the matrices $\mathbf{B}_1$, $\mathbf{B}_2$, $\mathbf{B}_3$, $\mathbf{B}_4$ and the vector $\mathbf{B}_5$ delineate the linear constraints (6.40)–(6.44) in the compact form (6.47).

Although the multi-parametric programming theory can be currently applied to the LP problem (6.46)–(6.47) to get (1) permissible regions of power transfers and (2) the cost function $z\left(\mathbf{P}_{\text{grid}}\right)$, the intertemporal constraints in (6.47) may introduce a heavy computational burden, as similarly found in Sect. 2.4 in Chap. 2. Although two temporal acceleration-oriented resource aggregation methods have been developed in Sects. 2.4.1 and 2.4.2, they still have certain employment limitations in this case. Particularly, the resource aggregation method in Sect. 2.4.1 accelerates the computation by estimating a multi-period permissible region using several lower-dimension polytopes across different time periods. Since time periods may be overlapped in different lower-dimension polytopes, the cost function $z\left(\mathbf{P}_{\text{grid}}\right)$ cannot be well estimated. As for the resource aggregation method in Sect. 2.4.2 which decomposes intertemporal constraints, it only works for intertemporal constraints from generators while the intertemporal constraints in (6.47) come from ES. Consequently, let's next introduce a new decomposition strategy to ease the employment of multi-parametric programming theory to the LP problem (6.46)–(6.47).

The introduced decomposition strategy is illustrated in Fig. 6.5. Based on the LP problem (6.46)–(6.47), the intertemporal characteristics are reflected by that the charging and discharging power across consecutive periods are coupled through SoC. Consequently, let's consider two scenarios – one is with large demand fluctuations

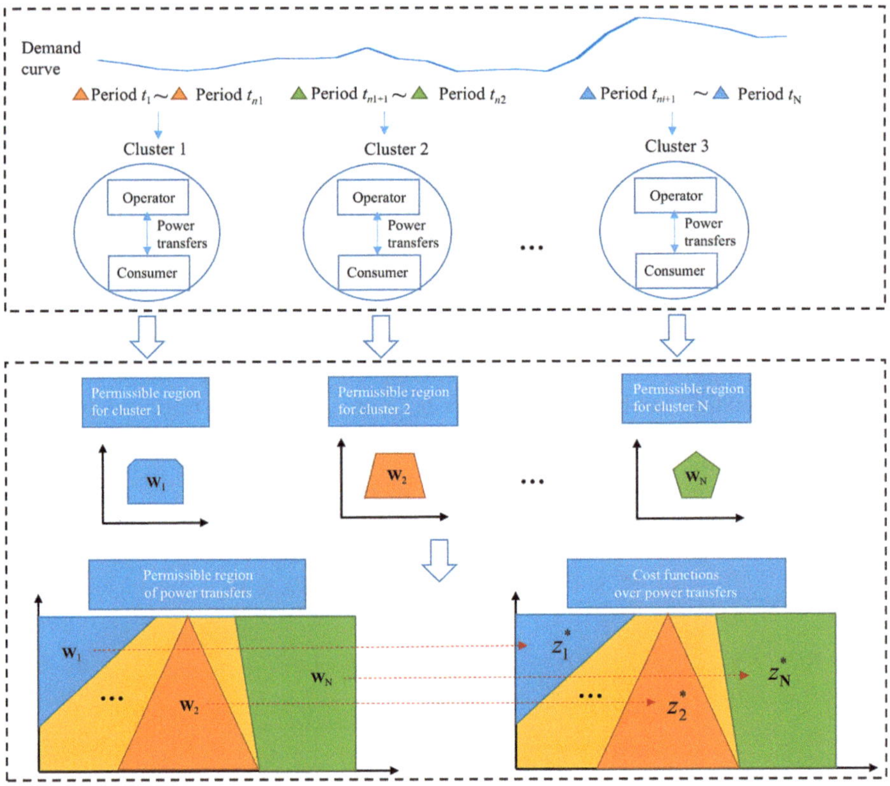

Fig. 6.5 Decomposition strategy, reprinted from [12], open access

while the other one is with small demand fluctuations—and their impacts on SoC. When demand fluctuations are large, the charging and discharging power may be also large, resulting in a wider range of SoC variations and weaker coupling across consecutive periods. When load fluctuations are small, the charging and discharging power is small, leading to a narrower range of SoC variations and stronger coupling across consecutive periods. The weak coupling relationships across consecutive periods can be approximately ignored. To achieve this goal, the cluster-based decomposition strategy is next implemented.

Let's first set the total number of cluster periods to $\mathrm{N}+1$. It takes the differences between consecutive data points of the daily demand curves, sorts these differences, and selects the indices of the top $\mathrm{N}+1$ differences, denoted as $\mathbf{K} = \{\mathrm{K}_1, \mathrm{K}_2, \ldots, \mathrm{K}_\mathrm{N}\}$. Based on these indices, the demands $\mathbf{P}_{\mathrm{load}}$ can be divided into $\mathrm{N}+1$ continuous blocks, represented as $(\mathbf{P}_{\mathrm{load},\mathrm{K}_i+1}, \ldots, \mathbf{P}_{\mathrm{load},\mathrm{K}_i})$, for $i = 1, 2, \ldots, \mathrm{N}$. These $\mathrm{N}+1$ continuous blocks lead to the decomposition strategy to ease the employment of multi-parametric programming to the LP problem (6.46) and (6.47). Suppose that the kth block in the LP problem (6.46) and (6.47) is expressed as:

$$\min_{\mathbf{x}_k, \mathbf{P}_{\text{grid},k}, \mathbf{P}_{\text{d},k}, \mathbf{P}_{\text{c},k}} z_k = \mathbf{A}_{1,k}^\top \mathbf{x} + \mathbf{A}_{2,k}^\top \left(\mathbf{P}_{\text{d},k} + \mathbf{P}_{\text{c},k}\right), \tag{6.48}$$

$$\text{s.t. } \mathbf{B}_{1,k}^\top \mathbf{x} + \mathbf{B}_{2,k}^\top \mathbf{P}_{\text{grid}} + \mathbf{B}_{3,k}^\top \mathbf{P}_{\text{d}} + \mathbf{B}_{4,k}^\top \mathbf{P}_{\text{c}} \le \mathbf{B}_{5,k}^\top. \tag{6.49}$$

By applying multi-parametric programming to the LP problem (6.48)–(6.49) for each block, the permissible region of power transfers from one power consumer can be estimated as $\cup_{k=1}^{\text{K}_\text{N}} \mathbf{W}_k$ and the cost function $z^* (\mathbf{P}_{\text{load}})$ can be estimated as $\sum_{k=1}^{\text{K}_\text{N}} z_k^* (\mathbf{P}_{\text{load},k})$, where $\mathbf{W}_k$ is the permissible region in the domain of $\mathbf{P}_{\text{load},k}$ by applying multi-parametric programming to the LP problem (6.48)–(6.49), and $z_k^* (\mathbf{P}_{\text{load},k})$ is the cost function by also applying multi-parametric programming to the LP problem (6.48)–(6.49).

6.4.2.2 Incorporation of Permissible Regions Into RODP

After each power consumer submits its permissible region $\cup_{k=1}^{\text{K}_\text{N}} \mathbf{W}_k$ and the cost function $\sum_{k=1}^{\text{K}_\text{N}} z_k^* (\mathbf{P}_{\text{load},k})$, the distribution operator can formulate the following RODP problem:

$$\min_{\mathbf{x}_{\text{DS}}, y_i, \mathbf{P}_{\text{load}}^{(i)}, \forall i} C_{\text{DS}} (\mathbf{x}_{\text{DS}}) + \sum_{i=1}^{\text{M}} \sum_{k=1}^{\text{N}_i} y_k^{(i)}, \tag{6.50}$$

$$\text{s.t. } \mathbf{C}_1^\top \mathbf{x}_{\text{DS}} + \sum_{i=1}^{\text{M}} \sum_{k=1}^{\text{N}_i} \mathbf{C}_{2,k}^{(i)^\top} \mathbf{P}_{\text{load},k}^{(i)} \le \mathbf{C}_3, \tag{6.51}$$

$$y_k^{(i)} \ge z_k^{(i)^*} \left(\mathbf{P}_{\text{load},k}^{(i)}\right), i = \{1, 2, \ldots, \text{M}\}, k = \{1, 2, \ldots, \text{N}_i\}, \tag{6.52}$$

$$\mathbf{P}_{\text{load},k}^{(i)} \in \mathbf{W}_k^{(i)}, i = \{1, 2, \ldots, \text{M}\}, k = \{1, 2, \ldots, \text{N}_i\}, \tag{6.53}$$

where $\mathbf{x}_{\text{DS}}$ stacks the decisions from the distribution operator, such as line, substation investments, and load shedding; the cost function C_{DS} reflects the investment and operation and maintenance costs by the distribution operator [11]; the superscript (i) indicates the ith power consumer; the constraints (6.50)–(6.53) delineate RODP requirements (e.g., branch flow limits, reliability requirements) and its mathematical formulation can be found in [11]; the constraints (6.52) reflect the cost function and permissible regions associated with power transfers from ES in power consumers, as similarly constructed in Sect. 6.2.2.1. Note that the reliability requirement in one distribution sub-system can be quantified by *expected energy not supplied* (EENS) [11], as also employed in this book.

6.4.3 *Case Studies*

The application method in Sect. 6.4.2 is implemented in a 24-bus distribution sub-system [14] and a 54-bus distribution sub-system [15], whose diagrams are given in Fig. 6.6. Their RODP results are given in Figs. 6.7 and 6.8 and Tables 6.4 and 6.5, with their comparison with those in the absence of power transfers from ES in power consumers.

Observed from Fig. 6.7 and Table 6.4 in the 24-bus distribution sub-system, the substation planning from the application method in Sect. 6.4.2 is the same as that without power transfers from ES in power consumers, while their line planning is different. Compared with the RODP results without power transfers from ES in power consumers, the application method in Sect. 6.4.2 does not invest lines 7–19, and 18–24 while building the line 1–14. This reduces the investment, operation, and maintenance costs by ¥3.04 $\times 10^6$, and reduces the total cost by ¥2.70 $\times 10^6$ when excluding the ES investment cost. This difference arises from that the application method in Sect. 6.4.2 can install ES instead of building lines for reliability improvements. It should be noted that the computational time of the application method in Sect. 6.4.2 is 13.48 h, while that without considering ES in power consumers is only 2.94. This is because the incorporation of permissible regions of power transfers from power consumers somehow increases the RODP complexity by adding additional constraints. Fortunately, the computational time may not be a concern in power industries, while the application method in Sect. 6.4.2 can exhibit better RODP results which ensure the same reliability with smaller costs.

Observed from Fig. 6.8 and Table 6.5 in the 54-bus distribution sub-system, the substation planning from the application method in Sect. 6.4.2 is the same as that without power transfers from ES in power consumers, while their line planning is different. Compared with the RODP results without power transfers from ES in power consumers, the application method in Sect. 6.4.2 does not build lines 21–54, 22–54, 23–24, 37–43, 40–41 while building the lines 8–33, 9–17, 31–37, 42–47. This reduces the investment, operation, and maintenance costs by ¥7.78 $\times 10^6$, and reduces the total cost by ¥6.06 $\times 10^6$ when excluding the ES investment cost. This difference arises from that the application method in Sect. 6.4.2 can install ES instead of building lines for reliability improvements.

6.5 Application Example 4. Heat and Electricity Integration

6.5.1 *Overview*

When one *power systems* (PS) and *heating systems* (HSs) are integrated through the *combined heat and power* (CHP) units, they can coordinate to optimize energy management and enhance overall system efficiency by utilizing the energy conversion

Fig. 6.6 Diagrams of the 24-bus and 54-bus distribution sub-systems, reprinted from [12], open access

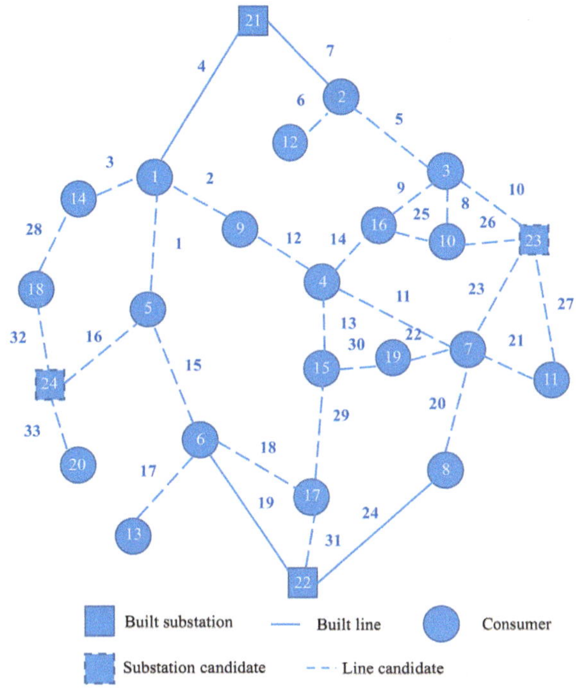

(a) Diagram of the 24-bus distribution sub-system

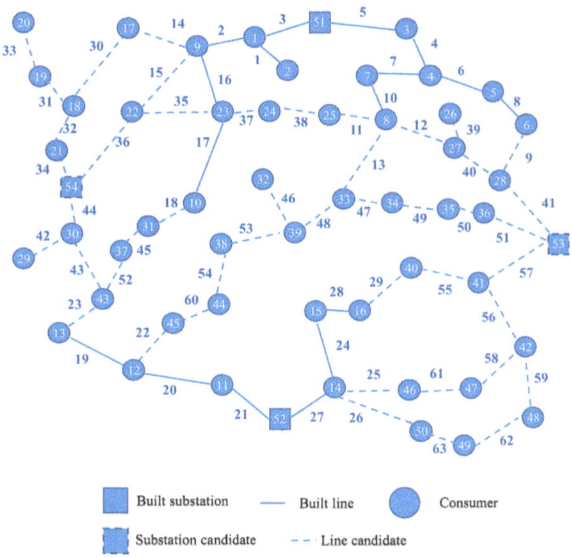

(b) Diagram of the 54-bus distribution sub-system

Fig. 6.7 RODP results under different methods in the 24-bus distribution sub-system, reprinted from [12], open access

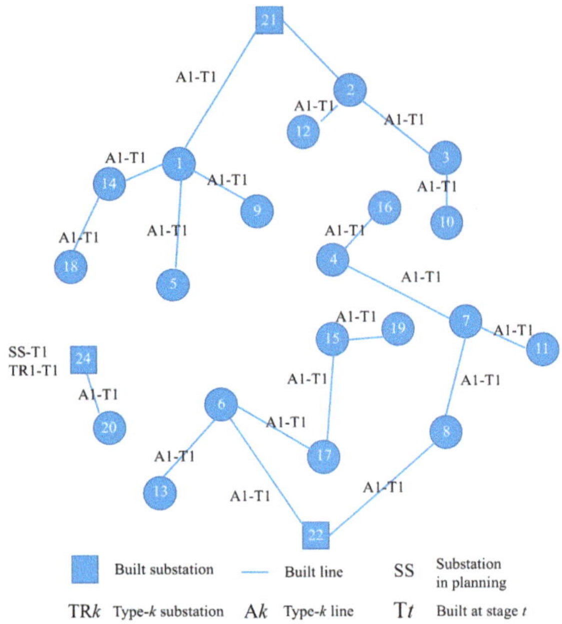

(a) RODP results from the method in Sec. 6.4.2

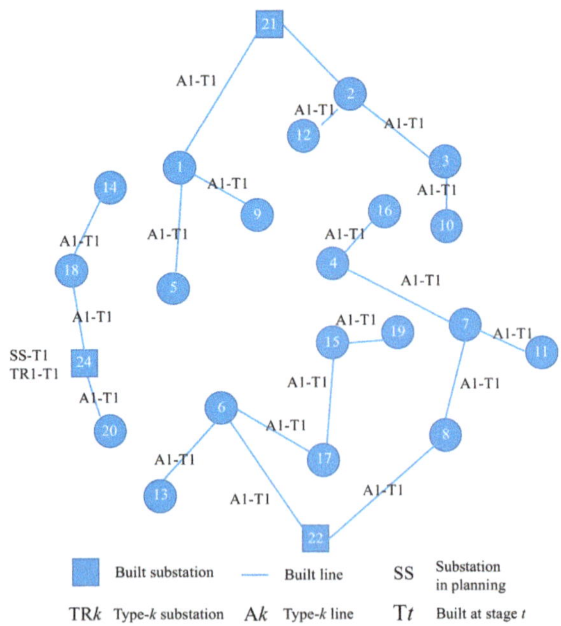

(b) RODP results without power transfers from power consumers

Fig. 6.8 RODP results under different methods in the 54-bus distribution sub-system, reprinted from [12], open access

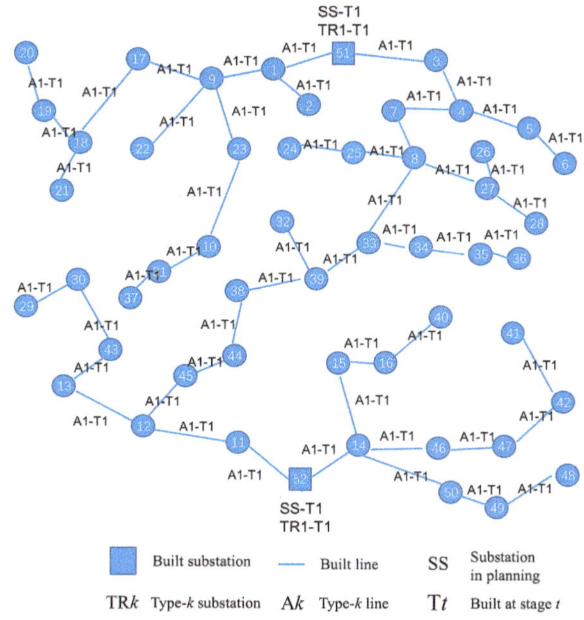

(a) RODP results from the method in Sec. 6.4.2

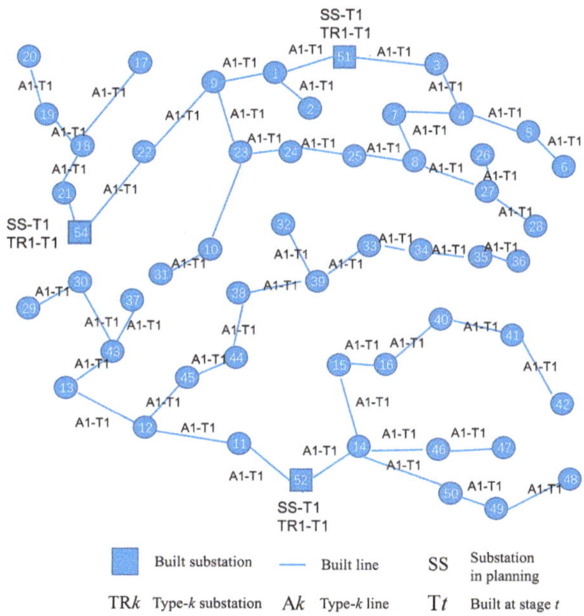

(b) RODP results without power transfers from power consumers

Table 6.4 RODP results under different methods in the 24-bus distribution sub-system, reprinted from [12], open access

Item	RODP results without power transfers from ES in power consumers	RODP results from the application method in Sect. 6.4.2
Total cost (¥)	1.43×10^7	1.16×10^7
Investment, operation and maintenance cost (¥)	1.29×10^7	9.86×10^6
EENS (MWh/year)	7.97	7.93
Reliability cost (¥)	1.53×10^5	1.47×10^5
ES investment cost (¥)	0	3.20×10^5
Computational time (h)	2.94	13.48

Table 6.5 RODP results under different methods in the 54-bus distribution sub-system, reprinted from [12], open access

Item	RODP results without power transfers from ES in power consumers	RODP results from the application method in Sect. 6.4.2
Total cost (¥)	1.54×10^7	9.34×10^6
Investment, operation and maintenance cost (¥)	1.52×10^7	7.42×10^6
EENS (MWh/year)	3.39	4.00
Reliability cost (¥)	2.17×10^5	2.56×10^5
ES investment cost (¥)	0	1.93×10^5
Computational time (h)	1.21	5.51

devices and the thermal dynamics of the HS. However, the implementation of the coordinated operation often requires access to comprehensive parameters from the other system, which can be challenging due to privacy and data protection concerns. Moreover, to address the HS model in *partial-differential algebraic equation* (PDAE) form, discretion-based numerical methods are commonly required. This introduces massive redundant variables and constraints and hinders real-time optimization in large-scale systems if the HS model is directly utilized. This section, alternatively, will employ permissible regions of the HS and CHP unit to address these problems. Due to the distinct inertia of thermal flow, the response speed of the HS is relatively slower. The requirement for optimal operation in the HS is not as stringent as in the PS, leading to relatively coarser management. Thus, the time intervals for operational optimization in the HS are relatively large, and only the static CHP unit model is adopted in Sect. 6.5.

6.5.2 Implementation Details

6.5.2.1 Permissible Region-Based Coordinated Operation

The ED problem for the integrated electricity and heating system can be formulated as

$$\min \sum_{t} \sum_{i \in \Theta_e \cup \theta_s} F_i \left(\mathbf{P}_{G,i,t}, \boldsymbol{\phi}_{chp,i,t}, \mathbf{P}_{\mathrm{chp},i,t} \right), \tag{6.54}$$

$$\text{s.t.} \quad \mathbf{A}_1 \leq \mathbf{f}_1 \left(\mathbf{P}_{\mathrm{chp},t}, \boldsymbol{\phi}_{\mathrm{chp},t} \right) \leq \boldsymbol{B}_1, \tag{6.55}$$

$$\mathbf{f}_2 \left(\mathbf{T}^{\mathrm{s}}_{\mathrm{nd},t}, \mathbf{T}^{\mathrm{r}}_{\mathrm{nd},t}, \boldsymbol{\phi}_{\mathrm{chp},t}, \mathbf{T}^{\mathrm{s}}_{t}, \mathbf{T}^{\mathrm{r}}_{t} \right) = \mathbf{0}, \tag{6.56}$$

$$\mathbf{A}_2 \leq \mathbf{f}_3 \left(\mathbf{T}^{\mathrm{s}}_{\mathrm{nd},t}, \mathbf{T}^{\mathrm{r}}_{\mathrm{nd},t} \right) \leq \mathbf{B}_2, \tag{6.57}$$

$$\mathbf{A}_3 \leq \mathbf{f}_4 \left(\mathbf{P}_{\mathrm{chp},t}, \mathbf{P}_{\mathrm{G},t} \right) \leq \mathbf{B}_3, \tag{6.58}$$

$$\mathbf{f}_5 \left(\mathbf{P}_{\mathrm{chp},t}, \mathbf{P}_{\mathrm{G},t}, \mathbf{U}_t, \boldsymbol{\theta}_t \right) = \mathbf{0}, \tag{6.59}$$

where F is the cost function of the generators and CHP units; Θ_e and Θ_h are the sets of electric buses and heating nodes, respectively; $\mathbf{P}_G$ is the active generation levels; $\mathbf{P}_{\mathrm{chp}}$ and $\boldsymbol{\phi}_{\mathrm{chp}}$ are the electric and thermal outputs of the CHP units, respectively; $\mathbf{f}_1$, $\mathbf{A}_1$ and $\mathbf{B}_1$ are constant elements to describe the constraints of the CHP units in Sect. 4.2.2.3; $\mathbf{T}^{\mathrm{s}}_{\mathrm{nd}}$ and $\mathbf{T}^{\mathrm{r}}_{\mathrm{nd}}$ and are the node temperatures and pipeline temperatures in the supply networks, respectively; $\mathbf{T}^{\mathrm{s}}$, $\mathbf{T}^{\mathrm{r}}$ are the node temperatures and pipeline temperatures in the return networks, respectively; $\mathbf{f}_2$ is compacts the equalities (4.93)–(4.95) for the HS during its resource aggregation process in Sect. 4.4.2; $\mathbf{f}_3$, $\mathbf{A}_2$ and $\mathbf{B}_2$ are the constant elements to describe the node temperature constraints from (4.37)–(4.38); $\boldsymbol{\theta}_t$ and $\mathbf{U}_t$ are the phase angles and voltage magnitudes, respectively; $\mathbf{f}_4$, $\mathbf{A}_3$ and $\mathbf{B}_3$ are the constant elements to describe the inequality constraints (e.g., branch flow limits from Chap. 2) of the PS; $\mathbf{f}_5$ is the function to describe the equality constraints (e.g., power balance constraints from Chap. 2) of the PS;

6.5.2.2 An Illustrative Case for Permissible Region of the HS

An illustrative case is provided to demonstrate the permissible region of one HS. The system structure is shown in Fig. 6.9. The transfer paths $\boldsymbol{\phi}_{\mathrm{h}}$, diversion ratio D, transfer coefficient J of the boundary conditions, and the transfer coefficient R of the transfer path are listed in the coming (6.60)–(6.62).

$$\Phi^{\mathrm{h}}_{14} = \{1, 2, 3\}, \ \Phi^{\mathrm{h}}_{15} = \{1, 5\}, \ \Phi^{\mathrm{h}}_{16} = \{1, 2, 5\}, \tag{6.60}$$

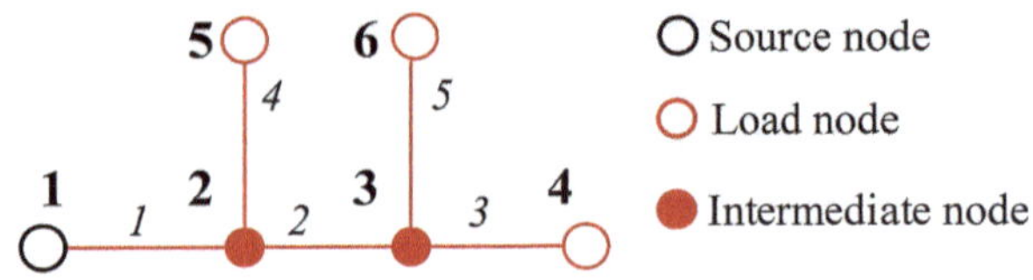

Fig. 6.9 Diagram of the illustrative HS, reprinted from [16], open access

$$\begin{cases} D_{14} = D_{15} = D_{16} = 1 \\ \xi_{14} = \xi_1 + \xi_2 + \xi_3 \\ \xi_{15} = \xi_1 + \xi_4 \\ \xi_{16} = \xi_1 + \xi_2 + \xi_5 \end{cases} \tag{6.61}$$

$$\begin{cases} R_{14} = J_{1,L_1} J_{2,L_2} J_{3,L_3} \\ R_{15} = J_{1,L_1} J_{4,L_4} \\ R_{16} = J_{1,L_1} J_{2,L_2} J_{5,L_5} \end{cases} \tag{6.62}$$

$$\begin{cases} T_{\mathrm{nd},4,t} = T^{\varphi}_{\mathrm{nd},4,t} + T^{\psi}_{\mathrm{nd},4,t} \\ T_{\mathrm{nd},5,t} = T^{\varphi}_{\mathrm{nd},5,t} + T^{\psi}_{\mathrm{nd},5,t} \\ T_{\mathrm{nd},6,t} = T^{\varphi}_{\mathrm{nd},6,t} + T^{\psi}_{\mathrm{nd},6,t} \end{cases}. \tag{6.63}$$

Based on (4.93), the temperature components of nodes 4–6 determined by boundary conditions are

$$\begin{cases} T^{\varphi}_{\mathrm{nd},4,t} = R_{14} T_{\mathrm{nd},1,t-\xi_{14}} \\ T^{\varphi}_{\mathrm{nd},5,t} = R_{15} T_{\mathrm{nd},1,t-\xi_{15}} \\ T^{\varphi}_{\mathrm{nd},6,t} = R_{16} T_{\mathrm{nd},1,t-\xi_{16}} \end{cases}, \tag{6.64}$$

Based on (4.94), the temperature components of nodes 4–6 determined by initial conditions are

$$\begin{aligned} T^{\psi}_{\mathrm{nd},4,t} = {} & \psi_1 \left(L_1 - v_1 \left(t - \xi_{24}\right)\right) J_1 \left(v_1 \left(t - \xi_{24}\right)\right) R_{24} \\ & + \psi_2 \left(L_2 - v_2 \left(t - \xi_{34}\right)\right) J_2 \left(v_2 \left(t - \xi_{34}\right)\right) R_{34}, \\ & + \psi_3 \left(L_3 - v_3 t\right) J_3 \left(v_3 t\right) \end{aligned} \tag{6.65}$$

$$\begin{aligned} T^{\psi}_{\mathrm{nd},5,t} = {} & \psi_1 \left(L_1 - v_1 \left(t - \xi_{25}\right)\right) J_1 \left(v_1 \left(t - \xi_{25}\right)\right) R_{25} \\ & + \psi_4 \left(L_4 - v_4 t\right) J_4 \left(v_4 t\right) \end{aligned}, \tag{6.66}$$

$$\begin{aligned} T^{\psi}_{6,t} = {} & \psi_1 \left(L_1 - v_1 \left(t - \xi_{26}\right)\right) J_1 \left(v_1 \left(t - \xi_{26}\right)\right) R_{26} \\ & + \psi_2 \left(L_2 - v_2 \left(t - \xi_{36}\right)\right) J_2 \left(v_2 \left(t - \xi_{36}\right)\right) R_{36}. \\ & + \psi_5 \left(L_3 - v_3 t\right) J_3 \left(v_3 t\right) \end{aligned} \tag{6.67}$$

By combining (6.64)–(6.67) with the node temperature constraints in (6.68), the permissible region of the HS can be built.

Fig. 6.10 Decision comparison between the methods M1 and M2 in Case I, reprinted from [17], with permission from Elsevier

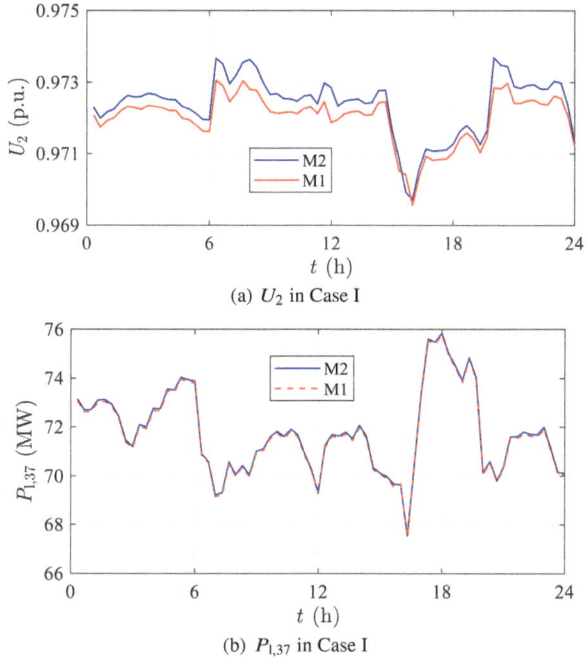

$$\begin{cases} T_{\mathrm{nd},4,\min} \le T_{\mathrm{nd},4,t} \le T_{\mathrm{nd},4,\max} \\ T_{\mathrm{nd},5,\min} \le T_{\mathrm{nd},5,t} \le T_{\mathrm{nd},5,\max} \\ T_{\mathrm{nd},6,\min} \le T_{\mathrm{nd},6,t} \le T_{\mathrm{nd},6,\max} \end{cases}. \tag{6.68}$$

6.5.3 *Case Studies*

The permissible region-based method (denoted as M1) is compared with the coordination method in a centralized manner (denoted as M2) through two cases. Case I contains one modified IEEE 118-bus PS and three independent HSs (with total 301 nodes) from [17]. The PS and HSs are coupled with three CHP units. Case II is an extension of Case II, where the modified IEEE 118-bus PS couples the seven HSs through seven CHP units (with 801-node HS). The system topologies and parameters can be found in [17]. The ED horizon is 24 h and the time interval is set to 20 min.

Taking the results in a centralized manner as the benchmarks, the results of the permissible region-based method are shown in Fig. 6.10 and Table 6.6, where (1) $\Delta T_{\max}$ is the maximum relative difference of the temperature, (2) $\Delta U_{\max}$ is the maximum relative difference of the voltage amplitude, and (3) $\Delta P_{\mathrm{G},\max}$ is the maximum relative difference of active generation levels.

From Table 6.6, although the voltage magnitude, thermal power, electric power, and temperature curves by the methods M1 and M2 present slight differences, the

Table 6.6 Comparison between the methods M1 and M2 in Case I, reprinted from [17], with permission from Elsevier

Method	$\Delta T_{\max}(\%)$	$\Delta U_{\max}(\%)$	$\Delta P_{G,\max}(\%)$	Cost ($)	Time (s)
M1	5.6	0.49	3.6	9236928	850
M2	–	–	–	9238737	990

Table 6.7 Constraint numbers of the methods M1 and M2 in Cases I and II, reprinted from [17], with permission from Elsevier

Method	Case I	Case II
M1	56645	76981
M2	78513	135201

operational costs of the methods M1 and M2 are almost the same. This indicates a desirable permissible region-based coordination performance. However, the computational time of the method M1 is smaller than that of the method M2. From Table 6.7, if there are (1) N_1 intermediate nodes, (2) N_2 source nodes, (3) N_3 load nodes, and (4) N_4 pipelines in the HS, the constraints of the method M2 include $3(N_2 + N_3)$ node temperature bounds and thermal power equations, N_2 return temperature equations at source nodes and N_3 supply temperature equations at load nodes. Since the permissible region simplifies the model constructions, the constraint number of the method M1 decreases, which presents the predominant practicality of the method M1 for a more efficient coordination.

Similar to that in Case I, the method M1 also vastly reduces the computational time over the centralized manner in Case II. Although there are slight differences in the coordination results (as shown in Fig. 6.11), the efficiency advantage of the method M1 increases with the system scales, which facilitate its practical implementations in the large-scale systems.

6.6 Summary

In this chapter, four application examples of employing permissible regions for coordinated energy management in low-carbon power systems are discussed.

The first application example demonstrates that the permissible regions of tie-line power flow exchanges can contribute to resource configurations by a tie-line power flow scheduling among transmission sub-systems even if different transmission sub-systems are operated by different operators in current power industries. Particularly, the permissible regions pave one promising way by which the coordination center can formulate an optimization problem without requiring all the detailed parameters in each transmission sub-system for a tie-line power flow exchange scheduling. The

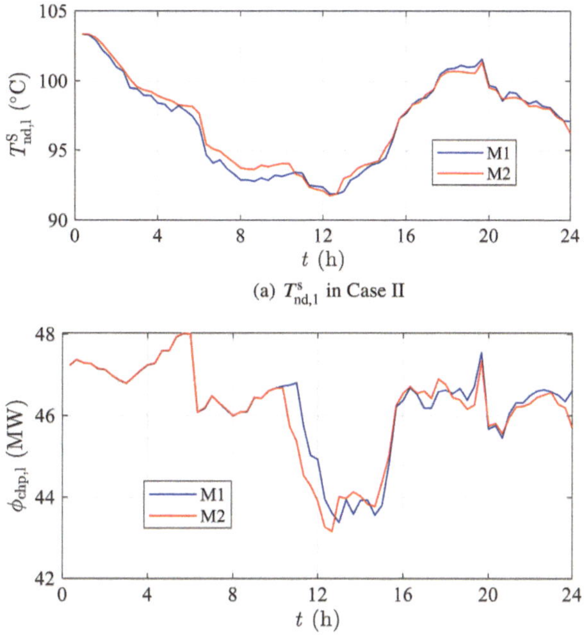

(a) $T^{s}_{nd,1}$ in Case II

(b) $\phi_{chp,1}$ in Case II

Fig. 6.11 Decision comparison between the methods M1 and M2 in Case II, reprinted from [17], with permission from Elsevier

resultant tie-line power flow exchange scheduling can be even the same as that in a centralized manner.

The second application example demonstrates that the permissible regions of PCC power flow exchanges can contribute to the coordination between one transmission sub-system and its interconnected distribution sub-systems when they are currently operated by different operators. Particularly, the permissible regions of PCC power flow exchanges can be incorporated into the transmission-level operation problem such that massive DERs can be utilized at the transmission level towards operations with lower costs when their detailed parameters are not required.

The third application example demonstrates that the permissible regions of power transfers from power consumers can contribute to the incorporation of ES in power consumers into RODP even if ES in power consumers is not invested by the distribution operator who conducts RODP. Particularly, when the reliability requirement holds, the permissible regions of power transfers from power consumers facilitate the cost savings arising from investing ES in power consumers instead of more expensive lines and substations by the distribution operator.

The fourth application example demonstrates the coordination between one PS and HSs. By representing the completely complex model of one HS as one simpler permissible region, the coordination efficiency can be improved when the coordination performance is almost affected (compared with the benchmark from the centralized manner). This exhibits a desirable potential that the permissible region-based coordination manner can sever as a promising manner for the interaction of one PS with heterogeneous energy flows.

6.7 Appendixes

6.7.1 *Equivalence Between the MINLP Problem (6.1)–(6.5) and the LP Problem (6.6)–(6.9)*

The equivalence between the MINLP problem (6.1)–(6.5) and the LP problem (6.6)–(6.9) is proven by showing that they have the same optimal objective values.

For the MINLP problem (6.1)–(6.5), its optimal objective value z^*_{MINLP} can be obtained by traversing all the critical regions, i.e.,

$$z^*_{\mathrm{MINLP}} = \min\left\{\left(z^*_1(\mathbf{w})\right)^*, \left(z^*_2(\mathbf{w})\right)^*, \ldots, \left(z^*_{\mathrm{N}}(\mathbf{w})\right)^*\right\}, \tag{6.69}$$

where $\min\{*\}$ indicates the minimum value of the set $*$; $\left(z^*_j(\mathbf{w})\right)^*$ is the optimal objective value in the jth critical region, i.e.,

$$\left(z^*_j(\mathbf{w})\right)^* = \min_{\mathbf{w}} z^*_j(\mathbf{w}), \tag{6.70}$$

$$\text{s.t.}\quad \mathbf{G}_{\mathrm{w},j}\mathbf{w} \le \mathbf{F}_{\mathrm{w},j}, \tag{6.71}$$

$$z^*_j(\mathbf{w}) = \mathbf{G}_{\mathrm{z},j}\mathbf{w} + \mathbf{F}_{\mathrm{z},j}. \tag{6.72}$$

For the LP problem (6.6)–(6.9), its idea to find the optimal solution can be illustrated in Fig. 6.12.

The solid line in Fig. 6.12 represents the piece-wise convex function $z^*(\mathbf{w})$ over the permissible region of tie-line power flow exchanges [13]. By introducing a continuous variable y, the constraints (6.7) extend the optimization space to the epigraph of the piece-wise convex function $z^*(\mathbf{w})$, which includes both the shadow area and the solid line in Fig. 6.12. Since the objective function (6.6) aims to minimize the variable y, and the epigraph of the piece-wise convex function is convex, the optimal solution of y will return to the solid line at optimum. In other words, the minimum value y^* is the minimum over all the critical regions, i.e.,

$$y^* = \min\left\{\left(z^*_1(\mathbf{w})\right)^*, \left(z^*_2(\mathbf{w})\right)^*, \ldots, \left(z^*_{\mathrm{N}}(\mathbf{w})\right)^*\right\}. \tag{6.73}$$

Combining (6.69) with (6.73) claims the same optimal objective values of the MINLP problem (6.1)–(6.5) and the LP problem (6.6)–(6.9), i.e., the LP problem (6.6)–(6.9) is equivalent to the MINLP problem (6.1)–(6.5).

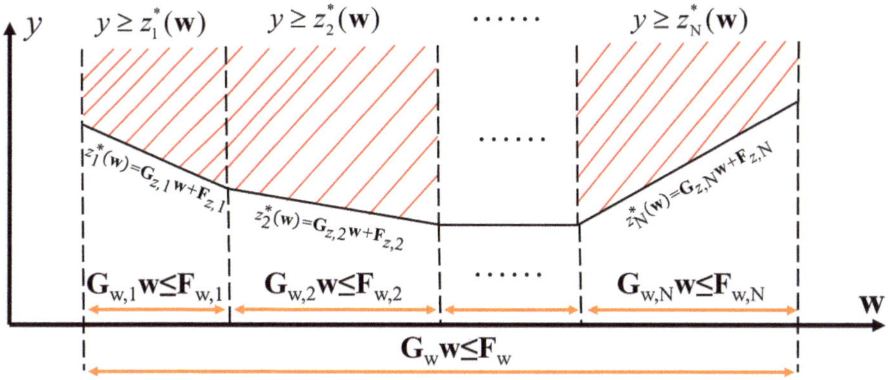

Fig. 6.12 Graphical illustration of the LP problem (6.6)–(6.9), reprinted from [18], with permission from Elsevier

6.7.2 Relaxing Complementarity Constraints of Energy Storage

The constraints of a general storage-concerned power system optimization reflect requirements on the charging power vector $\mathbf{z}_c$, discharging power vector $\mathbf{z}_d$, and the decision variable $\mathbf{x}$ (e.g., generation levels) except charging and discharging power. Such constraints over on $(\mathbf{x}, \mathbf{z}_c, \mathbf{z}_d)$ can be written in the following general form[1] with the binary variables $\mathbf{k}_c$ and $\mathbf{k}_d$:

$$(\mathbf{x}, \mathbf{z}_c + \mathbf{z}_d) \in \mathbf{C}, \tag{6.74}$$

$$\mathbf{x} \in \mathbf{X}, \tag{6.75}$$

$$\underline{\mathbf{S}} \leq \mathbf{S}_0 - \sum_{\tau=1}^{t} \left(\eta_c \otimes \mathbf{z}_{c,\tau} + \frac{1}{\eta_d} \otimes \mathbf{z}_{d,\tau} \right) \leq \overline{\mathbf{S}}, t = \{1, 2, \ldots, \mathrm{N_T}\}, \tag{6.76}$$

$$\mathbf{k}_c \otimes \underline{\mathbf{z}}_c \leq \mathbf{z}_c \leq \mathbf{0}, \mathbf{0} \leq \mathbf{z}_d \leq \mathbf{k}_d \otimes \overline{\mathbf{z}}_d, \tag{6.77}$$

$$\mathbf{k}^c + \mathbf{k}^d = \mathbf{1}, \tag{6.78}$$

where $\mathbf{C}$ and $\mathbf{X}$ are constraint sets; the operator $\otimes$ is the Hadamard product; $\underline{*}$ and $\overline{*}$ are the lower and upper bounds of the vector $*$, respectively; $\left(\mathbf{z}_{c,t}, \mathbf{z}_{d,t}\right)$ are the

[1] The non-convex complementarity constraints can be formulated in two equivalent forms: one is based on binary variables [19], and the other one is purely based on continuous variables [20]. This Appendix employs the former form for illustration purposes to demonstrate how to relax such non-convex complementarity constraints.

decisions of $(\mathbf{z}_c, \mathbf{z}_d)$ at time period t; η_c and η_d are charging and discharging efficiency coefficients, respectively.

The constraints (6.74) compact system-wide requirements (e.g., power balance) on the variables $(\mathbf{x}, \mathbf{z}_c, \mathbf{z}_d)$. The constraints (6.75) reflect the individual requirements on $\mathbf{x}$, such as generator capacity limits for $\mathbf{x}$. The constraints (6.76) represent the SoC limit at each time period t. The constraints (6.77) represent the upper and lower bounds of $(\mathbf{z}_c, \mathbf{z}_d)$. The constraints (6.78) are the complementarity constraints where the states of charging and discharging power are indexed by the binary variables $\mathbf{k}_c$ and $\mathbf{k}_d$, respectively. Note that (1) different power flow models can yield different forms of the set $\mathbf{C}$ and thus $\mathbf{C}$ may be a polyhedral, convex, or non-convex set, and (2) the vector $\mathbf{x}$ can include not only continuous variables but also integer variables.

The objective functions of a general storage-concerned power system optimization problem can be written in the following form:

$$\min_{\mathbf{x}, \mathbf{z}_c, \mathbf{z}_d, \mathbf{k}_c, \mathbf{k}_d} f(\mathbf{x}, \mathbf{z}_c, \mathbf{z}_d). \tag{6.79}$$

The objective function (6.79) can represent different goals by selecting different variables in an appropriate f. For example, the minimization power losses can be achieved by selecting $\mathbf{x}$ with the power loss function f, and the least-cost operation can be tackled by selecting $\mathbf{x}$, $\mathbf{z}_d$, and $\mathbf{z}_c$ with the cost function f.

The relaxation of the complementarity constraints stems from constructing a constraint set which is based on the original constraints (6.74)–(6.78). Particularly, the complementarity constraints (6.78) are replaced by the following linear constraints on $\mathbf{z}_c$.

$$\mathbf{S}_0 - \sum_{t=1}^{N_T} \left(\eta_c \circ \mathbf{z}_{c,t}\right) \leq \overline{\mathbf{S}}. \tag{6.80}$$

Also, the constraints (6.77) can be revised by directly restricting $(\mathbf{z}_c, \mathbf{z}_d)$ with their upper and lower bounds, i.e.,

$$\underline{\mathbf{z}}_c \leq \mathbf{z}_c \leq \mathbf{0}, \mathbf{0} \leq \mathbf{z}_d \leq \overline{\mathbf{z}}_d. \tag{6.81}$$

In summary, the relaxation can be achieved by constructing the following constraint set $\mathbf{Y}$:

$$\mathbf{Y} \triangleq \{(\mathbf{x}, \mathbf{z}_c, \mathbf{z}_d) \mid (6.74) - (6.76) \quad \text{and} \quad (6.80) - (6.81)\}. \tag{6.82}$$

Compared with the original constraints (6.74)–(6.78), the constraint set (6.82) does not include the binary variables $\mathbf{k}_c$ and $\mathbf{k}_d$ any more. Furthermore, the coming Proposition 6.1 shows how to get a solution feasible to (6.74)–(6.78) based on the constraint set (6.82).

Proposition 6.1 *The following strategy can recover any feasible point* $(\widetilde{\mathbf{x}}, \widetilde{\mathbf{z}}_c, \widetilde{\mathbf{z}}_d)$ *in (6.82) to a point* $(\check{\mathbf{x}}, \check{\mathbf{z}}_c, \check{\mathbf{z}}_d)$ *feasible to (6.74)–(6.78):*

$$\breve{\mathbf{x}} = \widetilde{\mathbf{x}}, \tag{6.83}$$

$$\breve{\mathbf{z}}_c = \widetilde{\mathbf{z}}_c + \{|\widetilde{\mathbf{z}}_c|, |\widetilde{\mathbf{z}}_d|\}_{\min}, \tag{6.84}$$

$$\breve{\mathbf{z}}_d = \widetilde{\mathbf{z}}_d - \{|\widetilde{\mathbf{z}}_c|, |\widetilde{\mathbf{z}}_d|\}_{\min}, \tag{6.85}$$

where the subscript "min" indicates the row-wise minimum element; $| * |$ is the absolute value of $*$. ■

Proof of Proposition 6.1. This proof is completed by showing the fulfillment of the original constraints (6.74)–(6.78) with $(\breve{\mathbf{x}}, \breve{\mathbf{z}}_c, \breve{\mathbf{z}}_d)$. Details are given below.

Since the strategy does not change the decisions on $\mathbf{x}$ and $(\mathbf{z}_c + \mathbf{z}_d)$ before and after recovery, the constraints (6.74)–(6.75) can be fulfilled by $(\breve{\mathbf{x}}, \breve{\mathbf{z}}_c, \breve{\mathbf{z}}_d)$. At the same time, (6.84)–(6.85) indicate that charge and discharge will not simultaneously occur within charging and discharging limitations, i.e. there is a pair $(\mathbf{k}_c, \mathbf{k}_d)$ such that the constraints (6.77)–(6.78) are fulfilled. This proof next focuses on the fulfillment of the SoC constraints (6.76).

(1) Fulfillment of the lower bounds of the SoC constraints (6.76). Based on (6.84)–(6.85), SoC at each time period t with the decisions $(\breve{\mathbf{z}}_c, \breve{\mathbf{z}}_d)$ can be formulated as

$$\mathbf{S}_0 - \sum_{\tau=1}^{t}\left(\eta_c \otimes \widetilde{\mathbf{z}}_{c,\tau} + \frac{1}{\eta_d} \otimes \widetilde{\mathbf{z}}_{d,\tau}\right) + \sum_{\tau=1}^{t}\left(-\eta_c + \frac{1}{\eta_d}\right) \otimes \{|\widetilde{\mathbf{z}}_c|, |\widetilde{\mathbf{z}}_d|\}_{\min}. \tag{6.86}$$

Since $\mathbf{S}_0 - \sum_{\tau=1}^{t}\left(\eta_c \otimes \widetilde{\mathbf{z}}_{c,\tau} + \frac{1}{\eta_d} \otimes \widetilde{\mathbf{z}}_{d,\tau}\right)$ is SoC at time period t with the decisions $(\widetilde{\mathbf{z}}_c, \widetilde{\mathbf{z}}_d)$, it fulfills the lower bound of the SoC constraints (6.76). Furthermore, the values of $\sum_{\tau=1}^{t}\left(-\eta_c + \frac{1}{\eta_d}\right) \otimes \{|\widetilde{\mathbf{z}}_c|, |\widetilde{\mathbf{z}}_d|\}_{\min}$ in (6.86) are non-negative since $\mathbf{0} \le \eta_c, \eta_d \le \mathbf{1}$. Consequently, SoC at each time period t with the decisions $(\breve{\mathbf{z}}_c, \breve{\mathbf{z}}_d)$ fulfills the lower bound of the SoC constraints (6.76).

(2) Fulfillment of the upper bounds of the SoC constraints (6.76). The worst case of violating the upper bound occurs when ES always charges. If the upper bound is not violated when ES is always charged at all the time periods, the upper bound at each time period t is always satisfied. The constraints (6.80) guarantee the fulfillment of the upper bounds under the worse case for the decisions $(\widetilde{\mathbf{z}}_c, \widetilde{\mathbf{z}}_d)$. Also, the charging power of $\widetilde{\mathbf{z}}_c$ is always less than that of $\breve{\mathbf{z}}_c$ based on the recovery rule (6.84). Consequently, the upper bounds of the SoC constraints (6.76) are fulfilled under the worse case for the decisions $(\breve{\mathbf{z}}_c, \breve{\mathbf{z}}_d)$. ■

Additionally, the optimal solutions of the optimization problem (6.74)–(6.79) can be obtained without the binary variables $\mathbf{k}_c$ and $\mathbf{k}_d$, as illustrated in the coming Proposition 6.2.

Proposition 6.2 *Suppose that the following two ex-ante conditions hold for the optimization problem (6.74)–(6.79):*

Condition 1. The function f in (6.79) is (1) non-decreasing in the domain of $\mathbf{z}_d$, and (2) non-increasing in the domain of $\mathbf{z}_c$.

Condition 2. The storage-side parameters fulfill the following algebraic inequalities:

$$\mathbf{S}_0 - \sum_{t=1}^{N_T} \left(\eta_c \otimes \underline{\mathbf{z}}_{c,t}\right) \leq \overline{\mathbf{S}}. \tag{6.87}$$

Given the conditions 1 and 2, the optimal solutions of the optimization problem (6.74)–(6.79) are also the optimal solutions of the following optimization problem without the binary variables $\mathbf{k}_c$ and $\mathbf{k}_d$.

$$v_{prop}^* = \min_{\mathbf{x},\mathbf{z}_c,\mathbf{z}_d} f\left(\mathbf{x}, \mathbf{z}_c, \mathbf{z}_d\right) \ s.t. \ (6.74) - (6.76) \ and \ (6.81). \tag{6.88}$$

■

Proof of Proposition 6.2. This proof is completed by showing that the optimal solutions of the optimization problem (6.88), which attains a lower bound of (6.74)–(6.78), is also a feasible solution of the optimization problem (6.74)–(6.78) under the given conditions 1 and 2. This indicates that the optimal solutions of the optimization problem (6.88) are indeed the optimal solutions of the optimization problem (6.74)–(6.78). Details are given below.

The constraints (6.74)–(6.78) are only a subset of the constraint set depicted by (6.74)–(6.76) and (6.81) in the domain of $(\mathbf{x}, \mathbf{z}_c, \mathbf{z}_d)$. Consequently, the optimal objective value v_{prop}^* of the optimization problem (6.88) is not larger than the optimal objective value v_{ori}^* of the original optimization problem (6.74)–(6.79).

The constraints (6.80) are redundant in the set $\mathbf{W}$ (6.82) when Condition 2 holds. Consequently, the optimal objective value v_{prop}^* of the optimization problem (6.88) equals the optimal objective value v_{W}^* of the following optimization problem:

$$v_{\text{W}}^* = \min_{\mathbf{x},\mathbf{z}_c,\mathbf{z}_d} f\left(\mathbf{x}, \mathbf{z}_c, \mathbf{z}_d\right) \text{ over } \mathbf{W}. \tag{6.89}$$

Moreover, Proposition 6.1 indicates that any feasible solution of the optimization problem (6.89) can be recovered to a feasible solution of (6.74)–(6.78). Particularly, Condition 1 allows that the objective value v_{rec}^* of f fixed by recovering the optimal solution of the optimization problem (6.89) is exactly v_{W}^*.

Since (1) v_{prop}^* is the lower bound of the optimization problem (6.74)–(6.78), 2) v_{prop}^* equals v_{W}^* and v_{W}^*equals v_{rec}^*, and 3) v_{rec}^* can be achieved at a feasible solution of the optimization problem (6.74)–(6.78), the optimization problem (6.74)–(6.78) can indeed attain its lower bound by solving the optimization problem (6.88). ■

References

1. Zhai, Q., Guan, X., Yang, J.: Fast unit commitment based on optimal linear approximation to nonliear fuel cost: error analysis and applications. Electr. Power Syst. Res. **79**(11), 1604–1613 (2009)
2. Yang, Z., Lin, W., Qiu, F., et al.: Correct the piecewise linearization error of quadratic objective function in DC OPF model. IEEE Trans. Power Syst. **34**(6), 5083–5086 (2019)
3. Lin, W.: Characterization and fast calculation of steady-state feasible region of tie-line transmission, Ph.D. thesis, Chongqing University (2021) (in Chinese)
4. Lin, W., Yang, Z., Yu, J., et al.: Improving security and economy of interconnected power network through explicit feasible region of tie-line power transfer. Int. J. Electr. Power Energy Syst. **123**, 106262 (2020)
5. Yang, Z., Zhong, H., Xia, Q., Kang, C.: Fundamental review of the OPF problem: challenges, solutions, and state-of-the-art algorithms. J. Energy Eng. **144**(1), 04017075 (2018)
6. VPP_AC. https://figshare.com/articles/dataset/VPP_AC/17082974
7. Lin, W., Yang, Z., Yu, J., et al.: Tie-line security region considering time coupling. IEEE Trans. Power Syst. **36**(2), 1274–1284 (2021)
8. Nguyen, H.D., Dvijotham, K., Turitsyn, K.: Constructing convex inner approximations of steady-state security regions. IEEE Trans. Power Syst. **34**(1), 257–267 (2019)
9. Lin, W., Zhao, C.: Improved characterization for AC-feasible power transfer regions of virtual power plants. Int. J. Electr. Power Energy Syst. **152**, 109260 (2023)
10. Li, Z., Wu, W., Tai, X., Zhang, B.: A reliability-constrained expansion planning model for mesh distribution networks. IEEE Trans. Power Syst. **36**(2), 948–960 (2021)
11. Hedayat, S., Reza, H., Mehdi, J.A.: Reliability improvement in radial electrical distribution network by optimal planning of energy storage systems. Energy **93**, 2299–2312 (2015)
12. Wang, Z.: A reliability-constrained distribution network expansion planning method considering flexibility space and benefit allocations of demand-side energy storages, Master Thesis, Chongqing University (2021) (in Chinese)
13. Pistikopoulos, E., Georgiadis, M., Dua, V.: Multi-parametric Programming: Theory, Algorithms and Applications. WileyVCH, Weinheim (2007)
14. Parameters of a 24-bus distribution network. https://doi.org/10.6084/m9.figshare.14822232.v5
15. Parameters of a 54-bus distribution network.https://doi.org/10.6084/m9.figshare.15241017.v3
16. Zhang, S., Gu, W., Lu, S., et al.: Integrated energy system analysis–from numerical to analytical (II): analytical method in district heating system. Proc. CSEE **44**(19), 7603–7615 (2024)
17. Zhang, S., Gu, W., Zhang, X., et al.: Fully analytical model of heating networks for integrated energy systems. Appl. Energy **327**, 120081 (2022)
18. Lin, W., Yang, Z., Yu, J., et al.: Determination of transfer capacity region of tie lines in electricity markets: theory and analysis. Appl. Energy **239**, 1441–1458 (2019)
19. Wang, Z., Zhong, J., Chen, D., et al.: A multi-period optimal power flow model including battery energy storage. In: 2013 IEEE Power & Energy Society General Meeting (2013)
20. Marley, J.F., Molzahn, D.K., Hiskens, I.A.: Solving multiperiod OPF problems using an AC-QP algorithm initialized with a SOCP relaxation. IEEE Trans. Power Syst. **32**(5), 3538–3548 (2017)

MIX
Papier aus verantwortungsvollen Quellen
Paper from responsible sources
FSC® C105338

If you have any concerns about our products,
you can contact us on
ProductSafety@springernature.com

In case Publisher is established outside the EU,
the EU authorized representative is:
Springer Nature Customer Service Center GmbH
Europaplatz 3, 69115 Heidelberg, Germany

Printed by Libri Plureos GmbH
in Hamburg, Germany